MATHIVATE

These mathlicious activities, projects, and games will put a <u>positive parabola</u> on everyone's face!!

MATH DAYS-OF-THE-WEEK

MATHOGGLE MONDAY = math game to boggle student's minds

TANGEO TUESDAY = fun with Tangrams

WUZZLE WEDNESDAY = math puzzles

THIRSTY THURSDAY = an "EquationANZA" of equations created from rolled dice to quench students' thirst for math

FOREMATHEAD FRIDAY = a participatory guessing game involving numbers or math vocabulary written on headbands

by Kim Thomas
2016 Illinois Teacher of the Year

ACKNOWLEDGMENTS

I want to start by thanking God for blessing me with a mathivating mind.
YOUR GRACE + GUIDANCE = MY INSPIRATION

This book is dedicated to:

My husband, Charlie, who is my constant source of love and laughter:
Thank you for sharing in this dream with me. I can't even put into numbers how much I ♥ you!!! I love x love + love ÷ love sharing my life with you!!

My children, Hank and Rosey:
Thank you for your understanding + support as I worked outside the home. Thank you for always making our home a fun x fun place!! Your determination to reach your dreams fills my heart with joy. I love you!!!

My bonus family, Andy, Ann, and Sophia:
I appreciate x appreciate your belief in me. Thank you for always listening to my ideas. Love you!!!

My parents Randy and Jane:
Thank you for your infinite x unconditional love. Love you 2,000 pounds!

My former and current students:
Thank you for inspiring me to be mathlicious. Every day you make me a better teacher. Love you all!

Teachers:
I know what you do and how difficult it can be. I exponentially thank you for being the length, width, and height for your students!!

Aric Bostick:
Thank you for all the encouragement + opportunities!! You found the speaker in me, and FIRED me UP to a whole new dimension!!

A huge thank you to Ruby Thompson, freelance designer & owner of Ruby Graphic Design, for the design, layout and editing expertise needed to make this resource available. I couldn't have done it alone!

Copyright © 2017
Kimberly Thomas

The purchase of this material entitles the buyer to reproduce worksheets and activities for classroom use only. Not for commercial resale or inclusion in any other printed or digital product or publication. Reproduction of these materials for an entire school or district is prohibited. No part of the book may be reproduced (except as noted above), stored in a retrieval system, or transmitted in any form or by any means (mechanically, electronically, recording, etc.) without the prior written consent of Kimberly Thomas.

Printed in the United States of America.

TABLE OF CONTENTS

INTRODUCTION	4
♥ 2 Know U	8
#MATH Muscles	10
CaMATHouflage	13
Combination Mathiplication	14
Me, Myself, and Math	16
Fracordiddle	18
NAMEREA (includes Templates)	20
MATH… It's in YOU!	36
MEflection SymMEtry	38
"IF I Had a FRACTION of Your Brain.."	40
Shop Til You Drop Scavenger Hunt	42
Circlize Yourself	46
Tantalizing Translations	48
Rockin' Reflections	50
TransINFORMations	52
Similar WHO? Similar YOU!	54
Name Your Game	56
Matherpieces	58
Always TIME for MATH!!	60
GEO in the Hood!	62
Checkbook Challenge	64
Mathlicious Book Club	68
MATHOLIDAY ACTIVITIES AND PROJECTS	71
Math'o'Lanterns	72
Happy MathO'ween	76
MaTHANKFULness…Gobble It Up!!	78
Gobbling Graphs	82
GEObread Houses	84
√144 Days of ChristMATH	90
Merry ChristMATH Trees	92
SnowMATH Flakegraphs	94
African American Mathematicians	96
MATHentines	100
MathEGGstatic Hunt	106
MATHLICIOUS GAMES	109
YahMATHzee!	110
Mathoggle!	114
EquationANZA	116
DynaMATHic Dice Dilemma	118
BaMATHanana	120
BinGRAPHo (includes Game Boards)	122
Brain Busting Boards	141
Mathlicious Jeopardy	142
OTHER MATHLICIOUS IDEAS	147

WARNING:

Implementing the use of the ideas in this book will cause children to Live, Laugh, Love, and LEARN math.

MATHLICIOUS

If you want your students to know that MATH does not stand for Makes All Things Horrible, please continue to read. The true meaning is: Makes All Things Happy or Happen or Hip. My vision for writing this book is that more students will live, laugh and love math!! Students will never want to leave math class. Yes, I said <u>never want to leave</u>. You probably think I am off my rocker, or on some wonderful medication that you wish you had. It can happen. When I finished my 23rd year of teaching, one of my journal questions was "What was the worst part of math class?" I had a student actually put….you guessed it… LEAVING. I was mathstatic!! The student who wrote it never showed much emotion. I never thought she enjoyed my class as much as she obviously did. My entire career has been spent teaching in an inner city school with awesome kids who come from low socioeconomic homes. I have completed most of these activities and projects with 6 x 5 students or sometimes a few more in a class. I am currently teaching in an alternative school for students who have been expelled. So, no excuses about large class sizes or behavior. It can be done – with a lot of fun!

My classroom mathagement style is keeping kids meaningfully engaged. I should mention that all of my activities and projects are completed in class. Yes, class time is used to complete projects. This allows you the privilege to monitor the learning and constructing process. Observing the empowerment is amathazing! These are NOT assigned as homework or take home. Take home projects are not a bonding experience for children and their parents. Trust me, I have been there. If a student insists on wanting to make up the project they missed while being absent, of course I allow them the opportunity. Ok, you may feel completely exhausted and want to hurt me afterwards, but your kids will want to hug you. I know it is hard sometimes, but remember you are there for EVERY student. Even the ones (because there is always more than one) that want to make you lose your religion. The contents in this mathlicious book are for every level of student. Teaching is NOT easy. It takes a lot of work and preparation to be effective and unforgettable. Unlike some students you may have that don't work hard, but yet are still unforgettable. Also, let your principal know that each activity is somehow related to the state standards. Administrators will be happy to know that while kids are having fun, they are still striving toward meeting the standards.

This book is straight to the point. Many of the activities can be completed daily and they never get bored with it. You will have to bear with me, I am most focused on sharing the spark of learning and telling you about the activities, so my tone is much more conversational than "grammatically correct".

When implementing this book and ideas, please remember to personalize your class project to your students. For example, I use every student's name at least one time during the school year on an activity sheet, journal, scratch your brain learning sheet, homework, or quiz. Many of my students have names that textbook companies have never used. They can't wait to see their name on the paper. Take time to get to know what your kids like.

The first day of school sets the tone for the entire year. When picking my students up, I ask them to please become linear and parallel. We will then translate ourselves to the classroom. Please remind them there is no rotating. Did you notice there are 4 math terms you can discuss with your students before they enter your room?!!! Yes they will think you are crazy. But that is okay. Most middle school math teachers are nuts. Kids need to know you are very serious about them learning as much as possible, and they can have fun. Here I go again using a word like FUN. Kids CAN learn and have fun. In all my years of teaching math, 85% or more of my students have met or EXCEEDED state standards. Just this past year, 29 out of my 90 students EXCEEDED on the state test. I don't say this to remind you of testing - I say this to reinforce that these projects I've used in my classroom for MANY years are effective in student growth, and that's why I want to share them with you!

Get to know your students! The first day of school, I pass my "♥ 2 know U" information sheet. Notice I ask them many questions that can't be found in the school's computer system. The kids enjoy this sheet much better than an index card. After I collect these, it is fun to read to the class what each student wrote, and then have them guess whose sheet you were reading!! Don't miss out on the opportunity to discuss all the great polygons. The reason I ask about their favorite cartoon is that I make sure I decorate an assessment with everyone's favorite cartoon throughout the year. Everything we do is math-related from day $\sqrt{1}$ until day $10^2 + 80$.

My license plate has CUN MATH on it. Say the letters then math. Yep you got it. "See You In Math". I have students create a license plate that represents them. I cut them out and hang them all over the classroom. Kids love to decorate the classroom and hallway with their artifacts.

So now for the mathlicious stuff. I wish you the **best ÷ best x best + best** year so far!!!

Cheer each other on the whole year through!!

MATHLICIOUS PROJECTS

Most projects are laid out with the project description & teacher's notes on the left side, and the reusable worksheet on the right side:

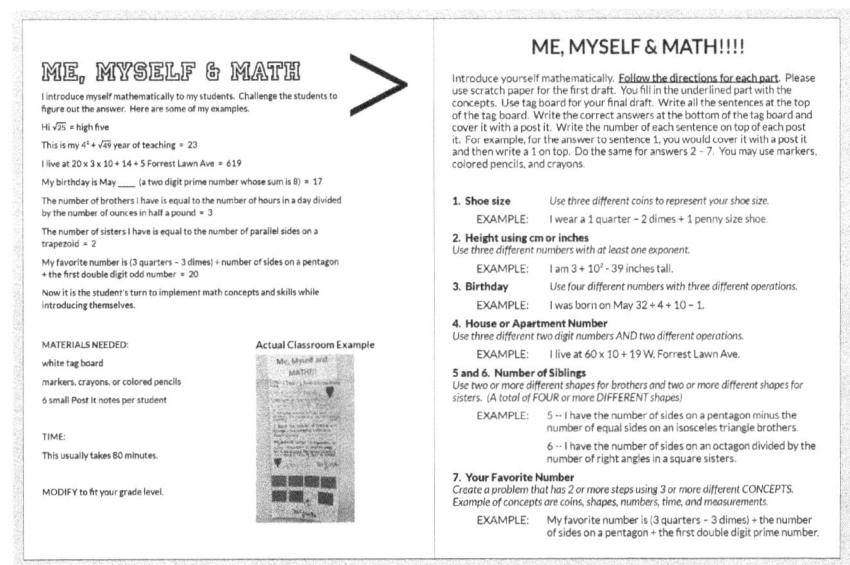

LOVE TO KNOW YOU

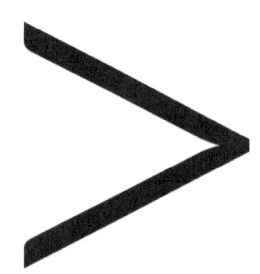

Get to know your students! The first day of school, I pass out this "♥ 2 know U" information sheet. Notice I ask them many questions that can't be found in the school's computer system. The kids enjoy this sheet much better than an index card. After I collect these, it is fun to read to the class what each student wrote, and then have them guess whose sheet you were reading!!

Don't miss out on the opportunity to discuss all the great polygons. The reason I ask about their favorite cartoon is that I make sure I decorate an assessment with everyone's favorite cartoon throughout the year. Everything we do is math-related from day $\sqrt{1}$ until day $10^2 + 80$.

♥ 2 Know U!!

Name: _____ is amathazing!!

$\sqrt{4}$ words that describe me are _____ and _____

The last school I attended was _____.

Please fill in the polygons:

Birthday

Favorite Candy

Favorite Song
(appropriate for school)

Favorite Television Show

What makes you laugh?

What are you going to learn to become?

Favorite Cartoon Character

Favorite Place To Eat

If you were given $100, what would you buy?

What about math is easy for you? _____

What about math is difficult for you? _____

MATH MUSCLES

I begin each day with Math Muscles on the board. These are five math questions that enhance the student's prior knowledge. I create them according to the content my students should have learned. These five questions a day equal to 25 questions a week which equals 900 questions a year!! These are so powerful in helping students retain the math being taught.

VERSION ONE: Each student receives a #MATH bingo card. They put the answers to the five questions wherever they choose (in the spaces on their card). You put answers on your own #MATH bingo card as well. Cut it up on Friday and put the answers in a bucket to pull for bingo. They will turn it in on Friday after the class plays several rounds of bingo.

VERSION TWO: Have students put the answers on paper labeled the day of the week and 1-5 (worksheet provided on following pages). On Fridays, I would put their names in a container and draw them out. I ask the question from that week's Math Muscles. If they had the correct answer they win a treat (or school incentive ticket), otherwise the treat value would go up for the next person I drew. They will turn their Math Muscles in on Friday after all the questions have been reviewed.

Example... Math Muscles Monday, Week 1

1. Evaluate $3x + y^2 - \sqrt{100}$ $x = 5$ $y = 4$

2. Round to the nearest hundredths place: 483.67521

3. If the area of a square is 100m², then what is the perimeter?

4. What is the product of the factors of 8?

5. Estimate the measure of this angle.

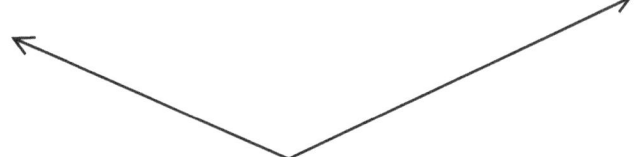

I always have at least one question that someone or some learning group can win a treat or a school incentive ticket. Like question 5. Whichever person or group estimates the closest wins. Kids LOVE it when they know they can get a reward. You can also give points to groups and then the group with the most points at the end of the week wins something.

M A T H

MUSCLES!!

Name:_____   Week:_____

MATH MUSCLES

Name _____   Week _____

Monday
1. _____
2. _____
3. _____
4. _____
5. _____

Tuesday
1. _____
2. _____
3. _____
4. _____
5. _____

Wednesday
1. _____
2. _____
3. _____
4. _____
5. _____

Thursday
1. _____
2. _____
3. _____
4. _____
5. _____

Friday
1. _____
2. _____
3. _____
4. _____
5. _____

caMATHouflage

This is another way I implement vocabulary. It is just what the word says. I camouflage a word with other letters on the board. I start with the long string of letters, erasing the letters one by one until a student shouts out the word.

(In the examples below, the underlined letters are the ones you know will remain, as letters are erased - remember you will not underline them when you write this on the board.)

Here are some examples:

    nmcueidbxpiarnocaglpsentr = circle

    vemdponerticdumrieshmgarpl = decimal

    fpcoairvmstoudiletemtaristn = formula

    wlhaierptnergcletsamhyirong = length

    hyemxpwioatidliagtpeintgor = mode

    osympcaeoltrsielcngteionlert = percent

    mipolysnirounallcemberispondt = pyramid

    rqdmduonalidbtmreuasinoetrl = quadrant

    porahnisolifigmtiarbotupsel = rhombus

    divrfoiurgalkeshoubliotngery = right

    rtselfcrvmatilsweellipnbtrae = scalene

Kids love to create these and share them with the class.

[This activity has no printable worksheet in the book]

LOCKER COMBINATION USING MATHIPLICATION

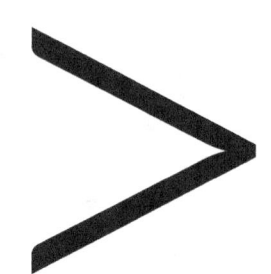

If you have students that use lockers, this is a mathawesome activity.

Combination Mathiplication

Name _____   Locker Combination _____

A. Directions: Follow the rule for each number in your locker combination.

Use **any** three numbers with **any** operations to equal the first number in your combination:

_____

Use three **different** numbers with two **different** operations to equal the second number in your combination:

_____

Use three **different** operations and four **different** numbers to equal the last number in your combination:

_____

B. For this section **NONE** of your facts can be duplicated.

What is the sum of your locker digits? _____

Write two facts about this number:

_____
_____

What is the difference of your first two digits? _____

Write two facts about this number:

_____
_____

What is the product of your two largest locker digits? _____

Write two facts about this number:

_____
_____

What is the quotient of the smallest and largest locker digits? _____

Write two facts about this number:

_____
_____

ME, MYSELF & MATH

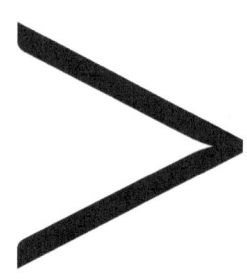

I introduce myself mathematically to my students. Challenge the students to figure out the answer. Here are some of my examples. You'll be inspired to create your own. Students LOVE to know about you!!!

Hi $\sqrt{25}$ = high five

This is my $4^2 + \sqrt{49}$ year of teaching = 23

I live at 60 x 10 + 19 W. Division Path = 619

My birthday is May ____ (a two digit prime number whose sum is 8) = 17

The number of brothers I have is equal to the number of hours in a day divided by the number of ounces in half a pound = 3

The number of sisters I have is equal to the number of parallel sides on a trapezoid = 2

My favorite number is (3 quarters – 3 dimes) ÷ number of sides on a pentagon + the first double digit odd number = 20

Now it is the student's turn to implement math concepts and skills while introducing themselves. If you have computers available, some students like to type these.

Actual Classroom Example:

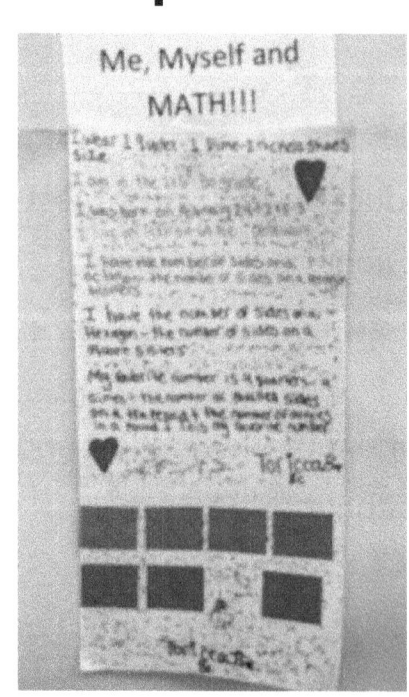

MATERIALS NEEDED:

white tag board

markers, crayons, or colored pencils

6 small Post It notes per student

TIME:

This usually takes 80 minutes.

Modify to fit your grade level.

ME, MYSELF & MATH!!!!

Introduce yourself mathematically. **Follow the directions for each part**. Please use scratch paper for the first draft. You fill in the underlined parts with the concepts that describe YOU. Use tag board for your final draft. Write all the sentences at the top of the tag board. Write the correct answers at the bottom of the tag board and cover it with a post it. Write the number of each sentence on top of each post it. For example, for the answer to sentence 1, you would cover it with a post it and then write a 1 on top. Do the same for answers 2 – 7. You may use markers, colored pencils, and crayons.

1. Shoe Size   *Use three different coins to represent your shoe size.*

    EXAMPLE: I wear a size <u>1 quarter – 2 dimes + 1 penny</u> shoe.

2. Height (using cm or inches)
Use three different numbers with at least one exponent.

    EXAMPLE: I am <u>$3 + 10^2 - 39$</u> inches tall.

3. Birthday   *Use four different numbers with three different operations.*

    EXAMPLE: I was born on May <u>$32 \div 4 + 10 - 1$</u>.

4. House or Apartment Number (doesn't have to be your real street)
Use three different two digit numbers AND two different operations.

    EXAMPLE: I live at <u>$60 \times 10 + 19$</u> W. Division Path

5 and 6. Number of Siblings
Use two or more different shapes for brothers and two or more different shapes for sisters. (A total of FOUR or more DIFFERENT shapes)

    EXAMPLE: 5 -- I have the <u>number of sides on a pentagon minus the number of equal sides on an isosceles triangle</u> brothers.

                         6 -- I have the <u>number of sides on an octagon divided by the number of right angles in a square</u> sisters.

7. Your Favorite Number
Create a problem that has 2 or more steps, using 3 or more different CONCEPTS. Examples of a concept are coins, shapes, numbers, time, and measurements.

    EXAMPLE: My favorite number is <u>(3 quarters – 3 dimes) ÷ the number of sides on a pentagon + the first double digit prime number</u>.

FRACORDIDDLE

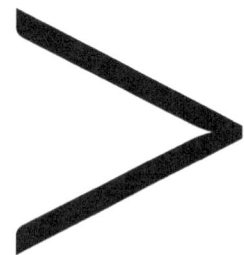

It is a funny word that kids love to say. I made this word up, and it is one way I introduce math terms. I ask my students what three words make up this word.

They are fraction, word, and riddle. If you notice FRAC is 1/2 of fraction, ORD is 3/4 of word, and IDDLE is 5/6 of riddle. A fracordiddle is a riddle using fractions of words. **The final word is always a math term**, but the words used to create it are not. This strengthens their vocabulary.

Here are some examples:

2/5 of n*ac*ho + 1/2 of p*ut*t + 1/3 of *e*lf = *acute*
1/2 of to*ad* + 1/2 of *jac*ket + 1/2 of d*ent*al = *adjacent*
2/5 of sm*al*l + 2/5 of lar*ge* + 3/8 of *bra*celet = *algebra*
2/5 of pl*an*t + 1/3 of lo*g* + 1/2 of *le*ft = *angle*
1/2 of l*av*a + 3/5 of op*era* + 2/3 of *ge*m = *average*
2/5 of por*ch* + 2/5 of sh*ar*k + 1/3 of *t*oe = *chart*
2/5 of *ch*ili + 1/3 of m*o*p + 1/2 of ne*rd* = *chord*
2/5 of fa*nc*y + 1/2 of *vi*olin + 1/2 of spi*der* = *cylinder*
2/5 of *ac*orn + 1/3 of u*mp*ire + 1/2 of sc*are*d = *compare*
3/7 of *com*welcome + 1/2 of *hip*pos + 3/5 of *sui*te = *composite*
3/5 of *con*bacon + 1/2 of *gr*umpy + 3/5 of sp*ent* = *congruent*
2/5 of sp*ad*e + 1/3 of *ci*rcus + 3/4 of *mal*t = *decimal*
3/5 of *dec*oy + 1/3 of *pr*etty + 3/4 of v*ase* = *decrease*
1/2 of *di*et + 1/2 of *ag*aag + 1/2 of n*eo*n + 1/2 of p*al*m = *diagonal*
1/2 of *di*ce + 2/5 of s*ta*mp + 1/2 of *fe*et + 2/3 of *er*a = *diameter*
1/3 of *e*beg + 1/3 of l*iq*uid + 1/2 of p*al*m = *equal*
1/2 of a*es* + 3/7 of s*tim*uli + 1/2 of h*ea*ter = *estimate*
1/3 of r*e*flex + 3/5 of *spon*ge + 1/2 of ab*sent* = *exponent*
1/2 of a*fra*id + 1/3 of i*ns*ect + 1/2 of b*io*nic = *fraction*
2/5 of a*gr*ee + 1/3 of f*a*n + 1/3 of ne*ph*ew = *graph*
1/2 of *chun*ky + 2/5 of *dr*ive + 1/2 of i*ce*d = *hundred*
1/2 of *ki*nd + 1/2 of *slo*gan + 3/5 of *fra*me = *kilogram*
1/2 of *ga*me + 2/3 of *gas* + 1/2 of *in*jure = *measure*
1/2 of *me*lt + 2/5 of *ra*dio + 1/2 of *ca*ne = *median*
2/5 of *ho*ney + 1/3 of a*lli*gator + 1/2 of s*tri*ve = *negative*
1/3 of f*a*n + 1/2 of *du*mp + 3/5 of *ber*ry = *number*
1/2 of *pa*ge + 1/2 of *ra*ce + 2/5 of *jol*ly + 1/2 of *se*lf = *parallel*
1/3 of *imp*atient + 2/5 of *pla*te + 2/7 of fur*na*ce = *pattern*
1/2 of *pen*cil + 1/2 of *out*age + 2/5 of *mon*th = *pentagon*
2/5 of *po*wer + 1/3 of *ly*rics + 2/3 of *ego* + 1/3 of *n*et = *polygon*
1/3 of *re*port + 2/5 of *sil*ly + 1/2 of *ti*me + 1/2 of *lo*ve = *positive*

Kids love to create these and share them with the class.

FRACORDIDDLE

Name_____

Directions: Create a FRACORDIDDLE that represents each description. Write the math term on the line and the fracordiddle in the space underneath. The words used to create your fracordiddle cannot be math terms. The letters used that create the math term must be adjacent (next to) each other. Write an explanation or give an example that proves you understand the meaning of each term.

Math Term

1. Use parts of **three** different words to create a math term: _____

Explanation and/or example_____
_____

2. Use parts of **three** different words to create a math term: _____

Explanation and/or example_____
_____

3. Use parts of **four** different words to create a math term: _____

Explanation and/or example _____
_____

4. Use parts of **four** different words to create a math term: _____

Explanation and/or example _____
_____

NAMEREA

Students enjoy showcasing their names. I use the letters of their name to spell MATH in the windows, the room, and/or the hallway. Students apply their measuring skills, and compute area and perimeter of composite figures.

Modify to fit your grade level.

Actual Classroom Example:

MATERIALS NEEDED:

A variety of tag board, poster board, or colored paper (tag board and poster board work best)

A set of block letters per student.

Ruler

TEMPLATES:

Following the worksheet on the next page, sketched templates are provided for each letter of the alphabet as a resource to you.

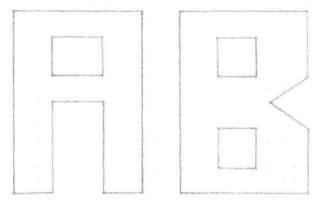

TIME:

This takes around three 80 minute class periods.

NAMEREA

Name: _____

Directions: Choose four letters from your name. List below (1,2,3,4 on this sheet). They must be different.

For 1 and 2, list the shapes that create the letter of your name next to the letter.

Using centimeters, measure the necessary dimensions of each shape on the template to compute the area of each letter. Make lines on your letters to break it up into shapes so you can compute area. Find the area of each section. Record the measurements used to find area on this sheet.

Letter in Your Name	Shapes in the Letter	Measure of Dimensions	Area
1. ____	_____	_____	_____
	_____	_____	_____
	_____	_____	_____
	_____	_____	_____
	_____	_____	_____
		Total Area of Letter:	_____
2. ____	_____	_____	_____
	_____	_____	_____
	_____	_____	_____
	_____	_____	_____
	_____	_____	_____
	_____	_____	_____
		Total Area of Letter:	_____

For letters 3 and 4, answer each question about your letter:

3. Write the letter you chose: _____   4. Write the letter you chose: _____

How many sides does the letter have? _____   How many sides does the letter have? _____

What is the name of the figure?_____   What is the name of the figure?_____

What are the measurements of each side?:   What are the measurements of each side?:

_____   _____

Sketch a drawing of your letter: Sketch a drawing of your letter:

What is the perimeter of your letter?_____   What is the perimeter of your letter?_____

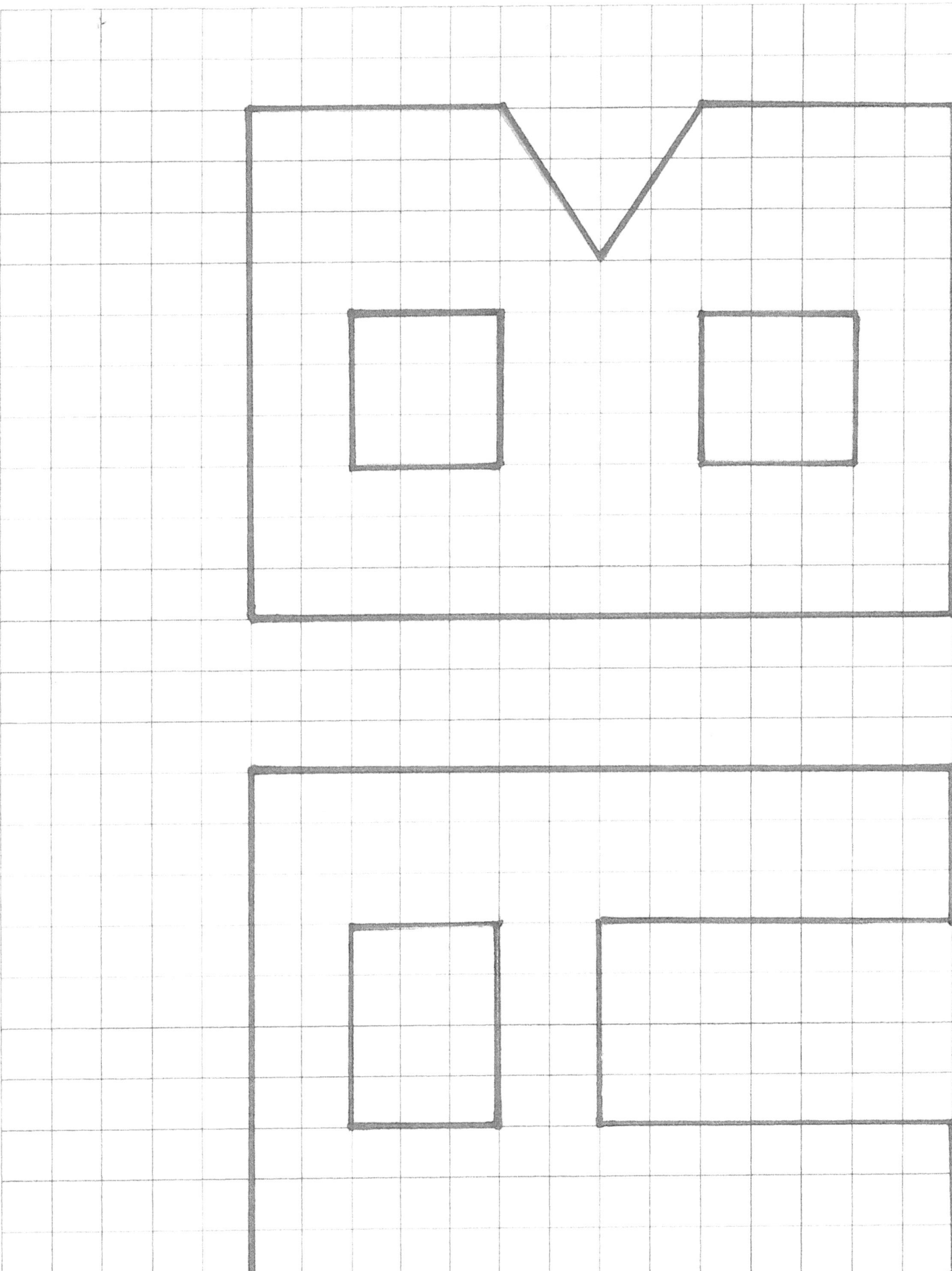

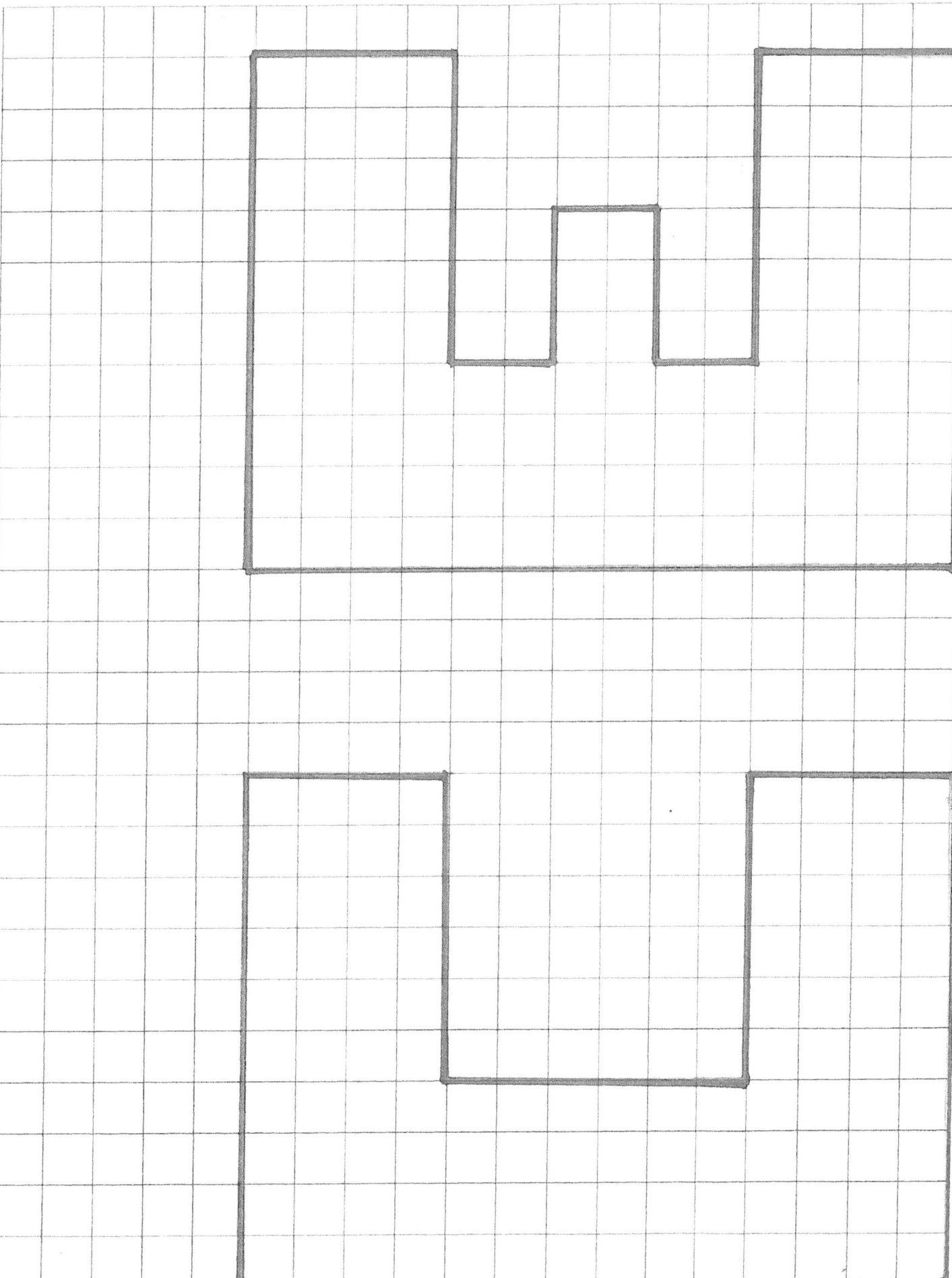

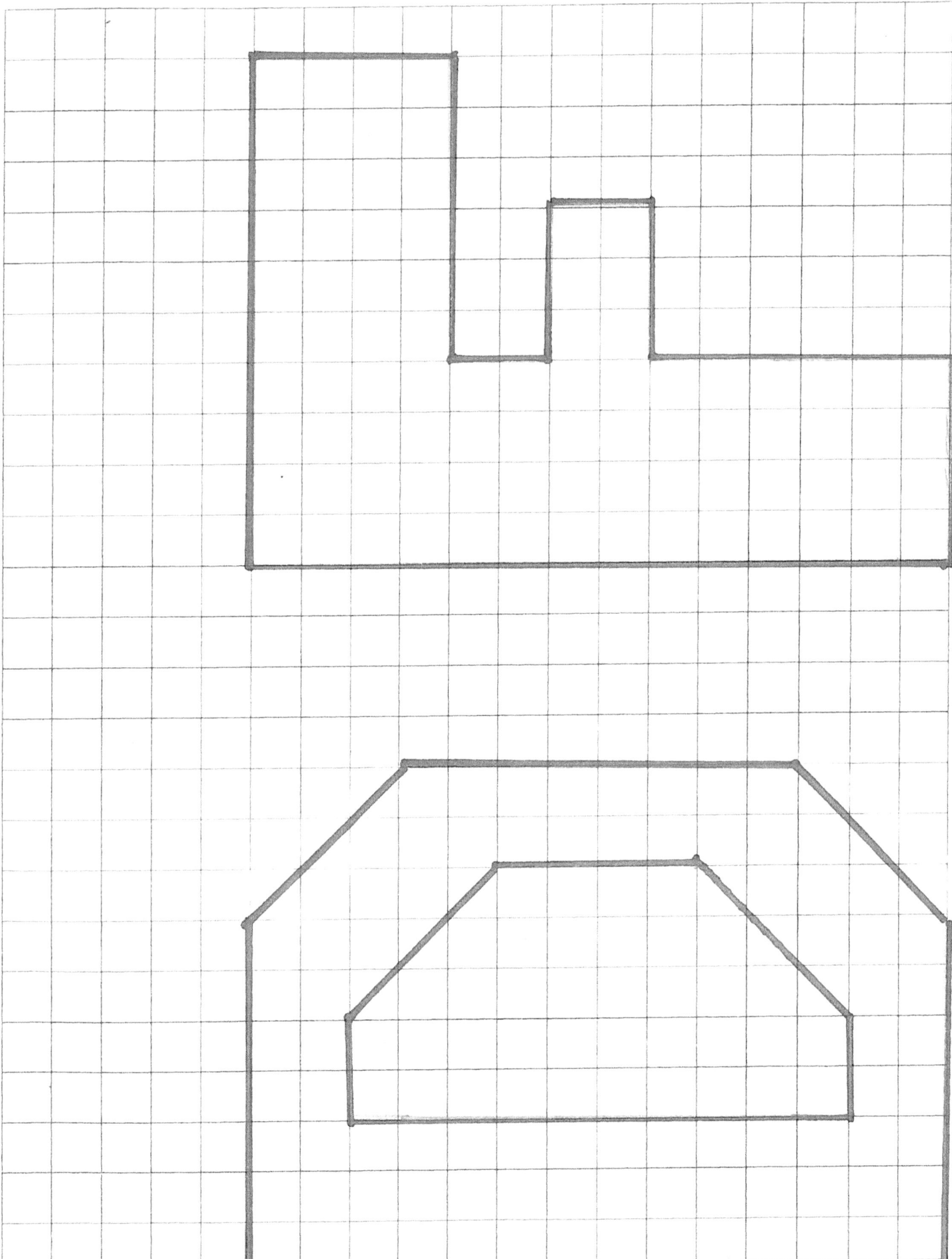

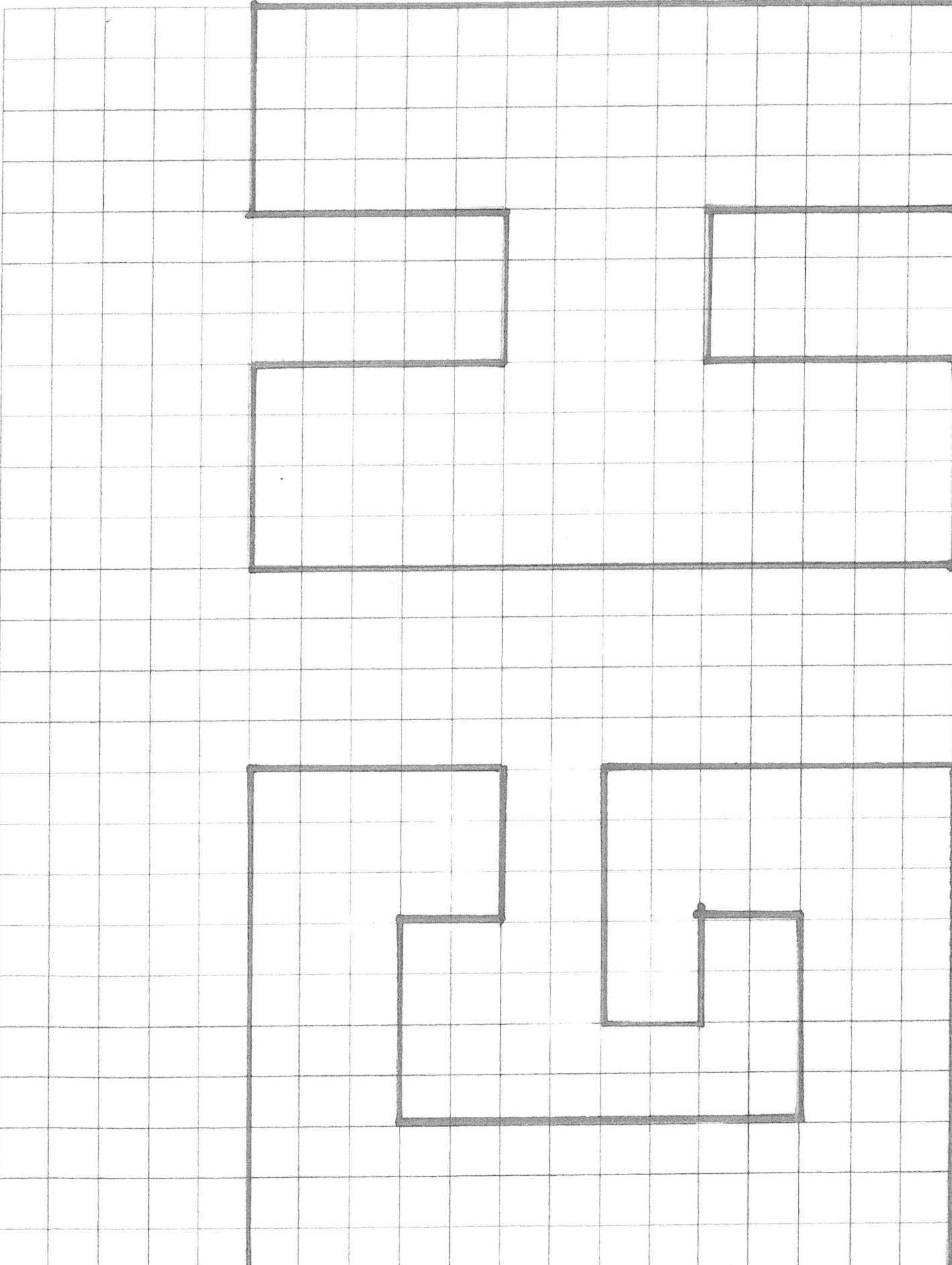

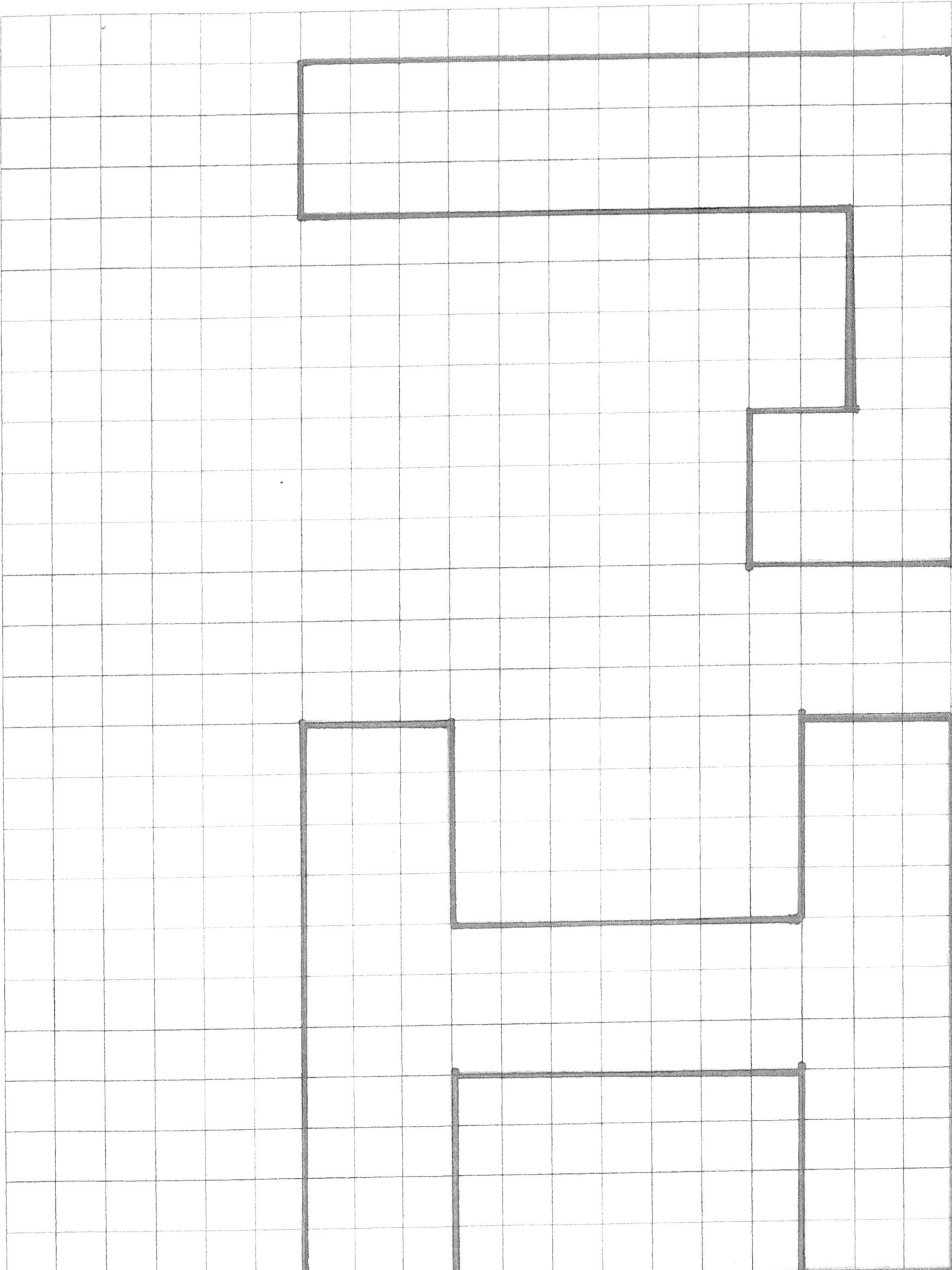

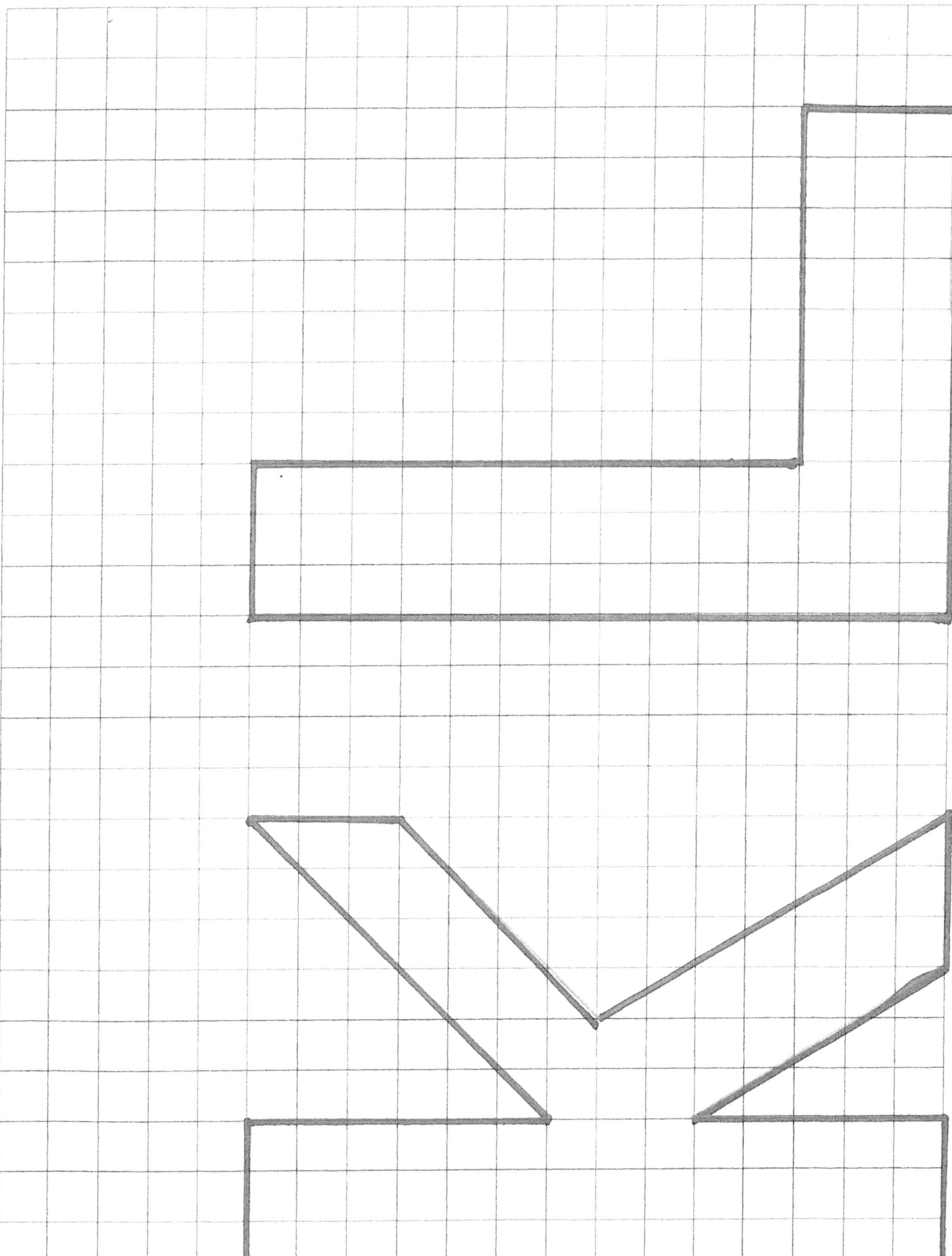

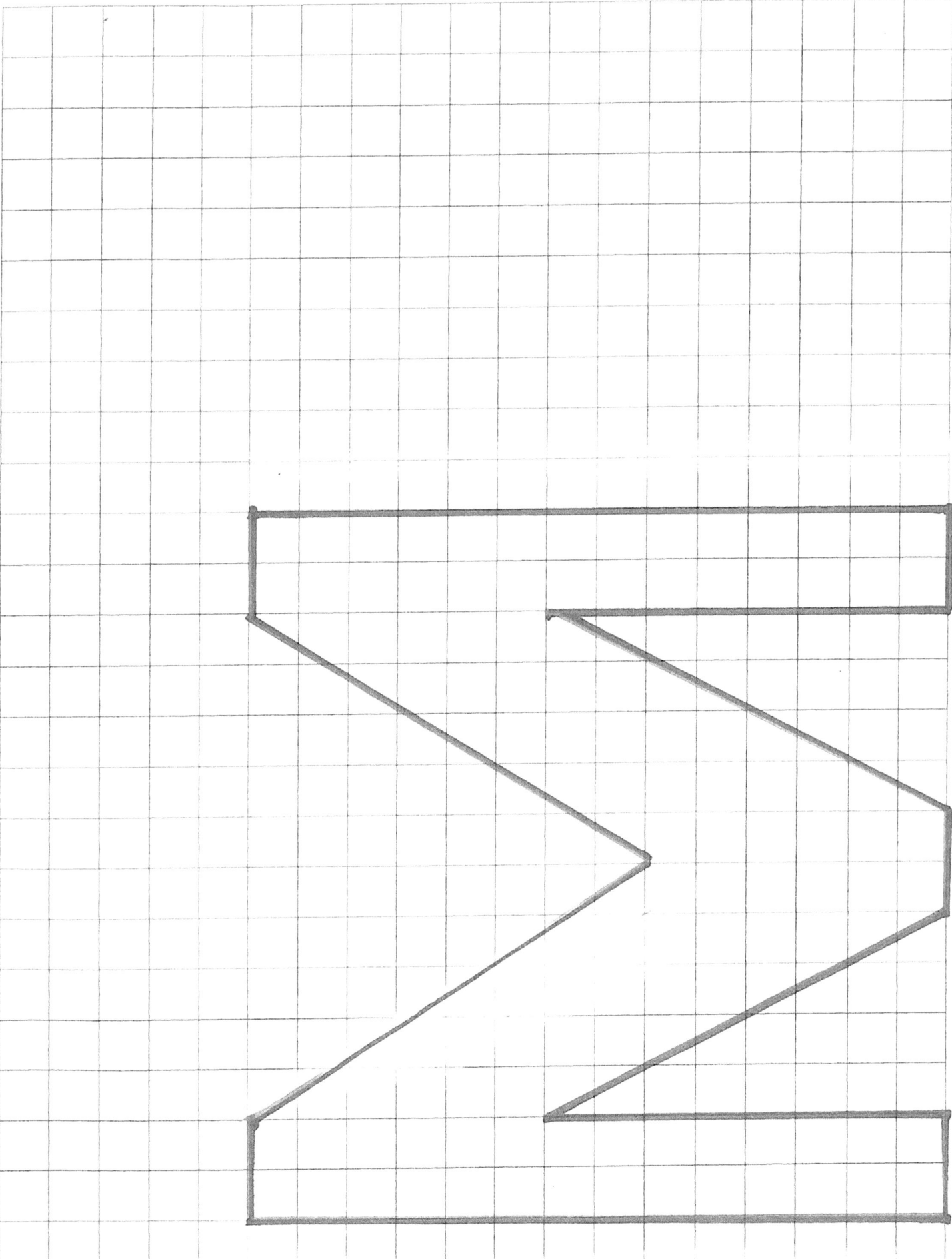

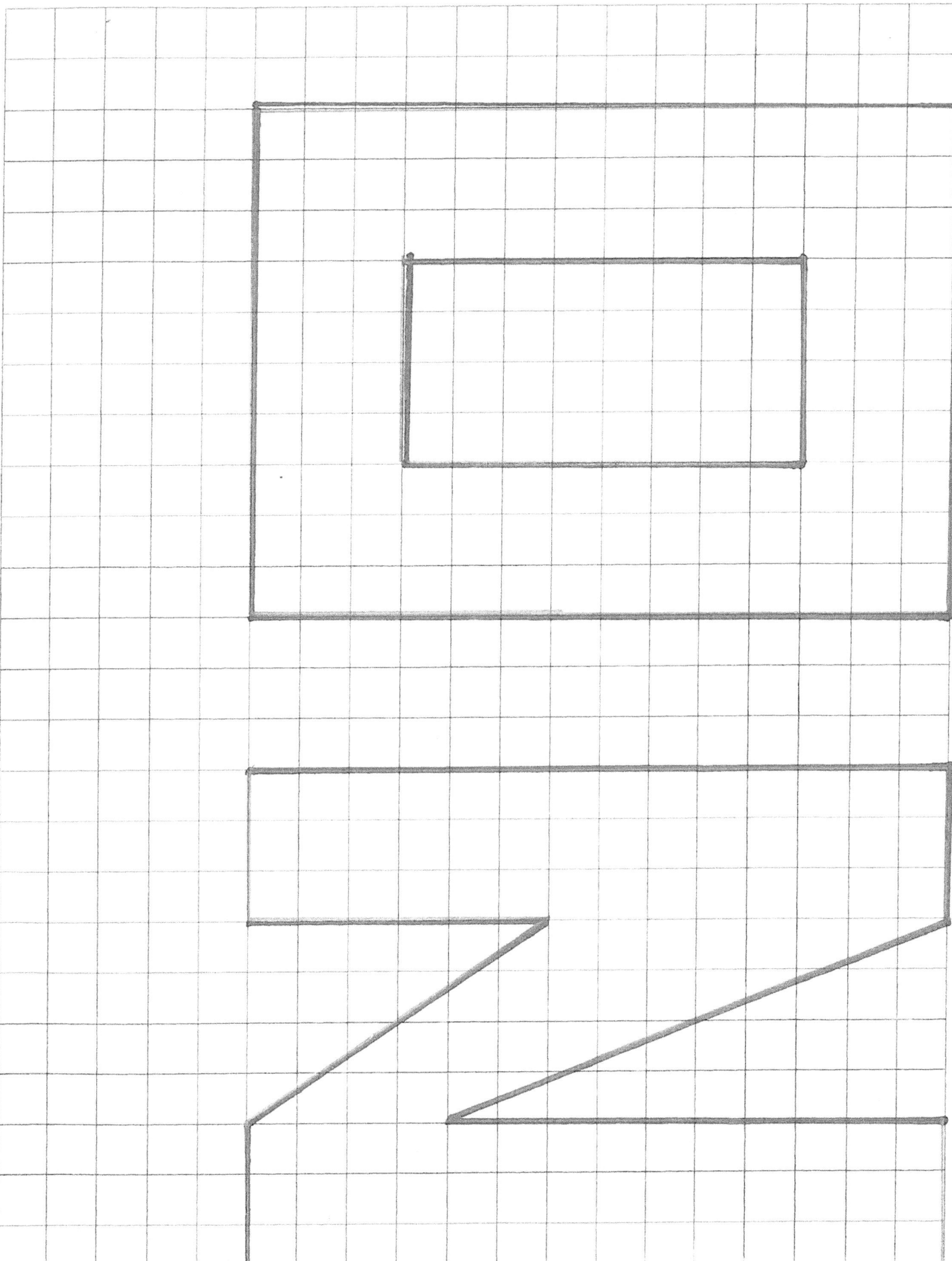

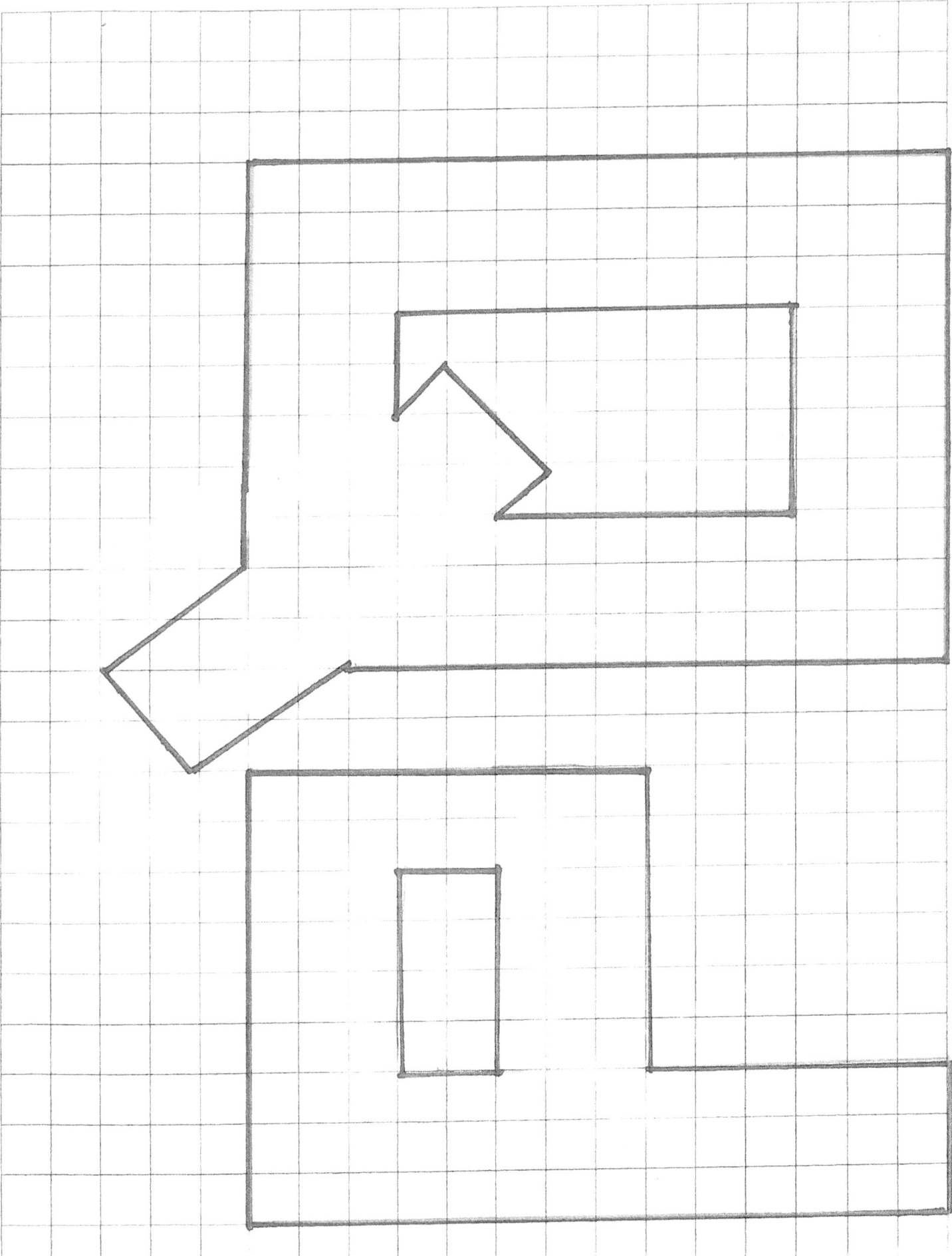

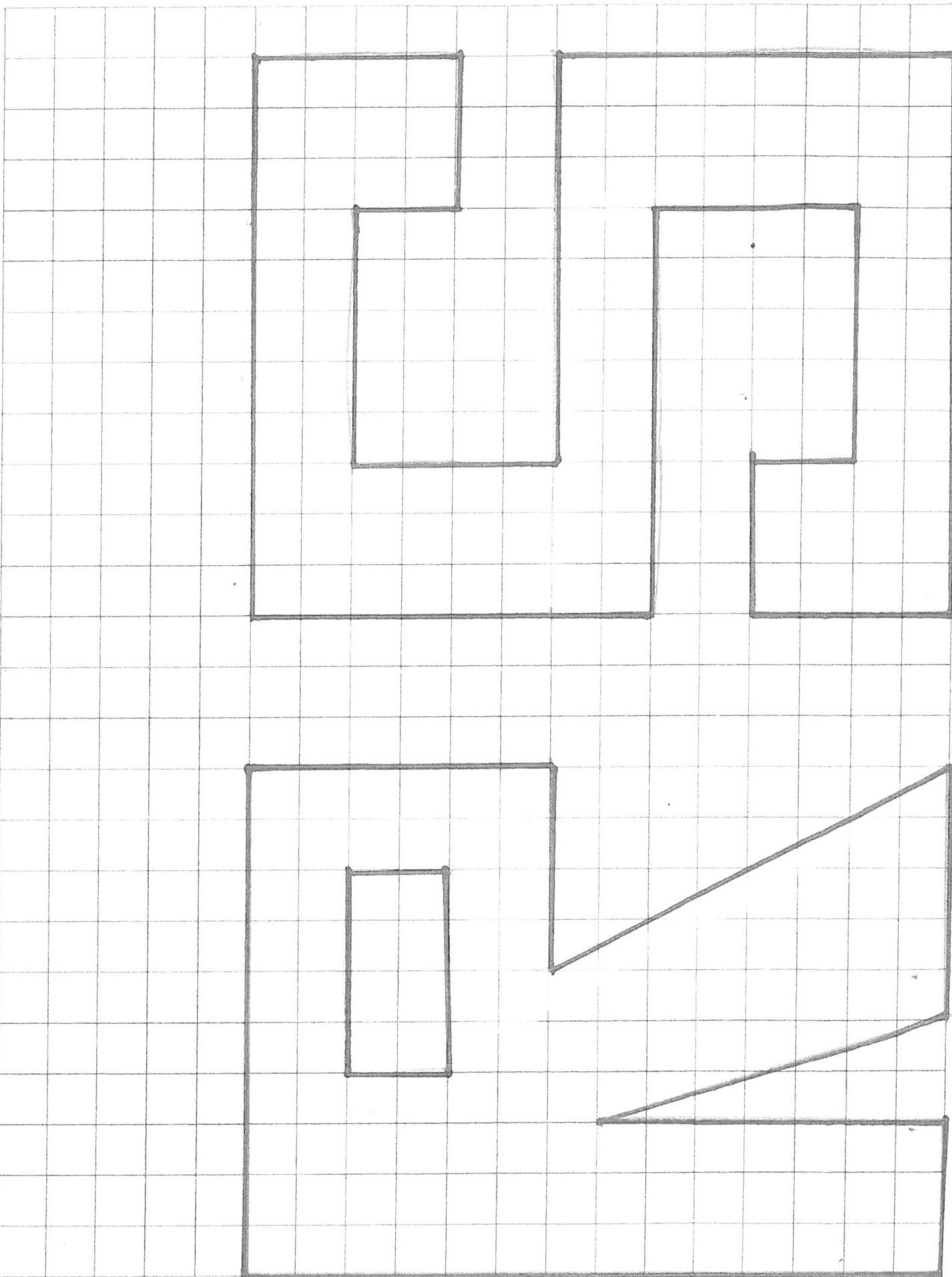

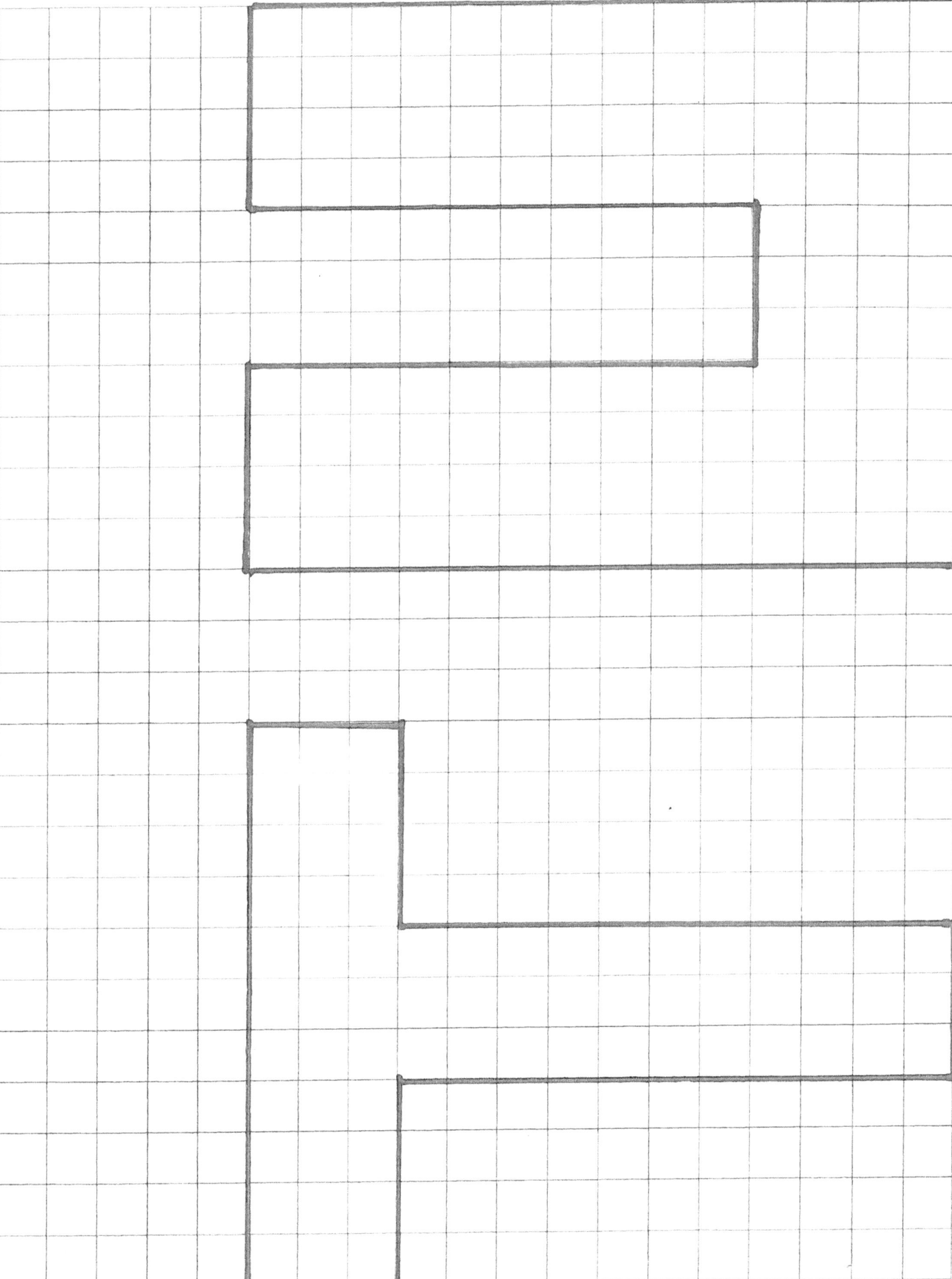

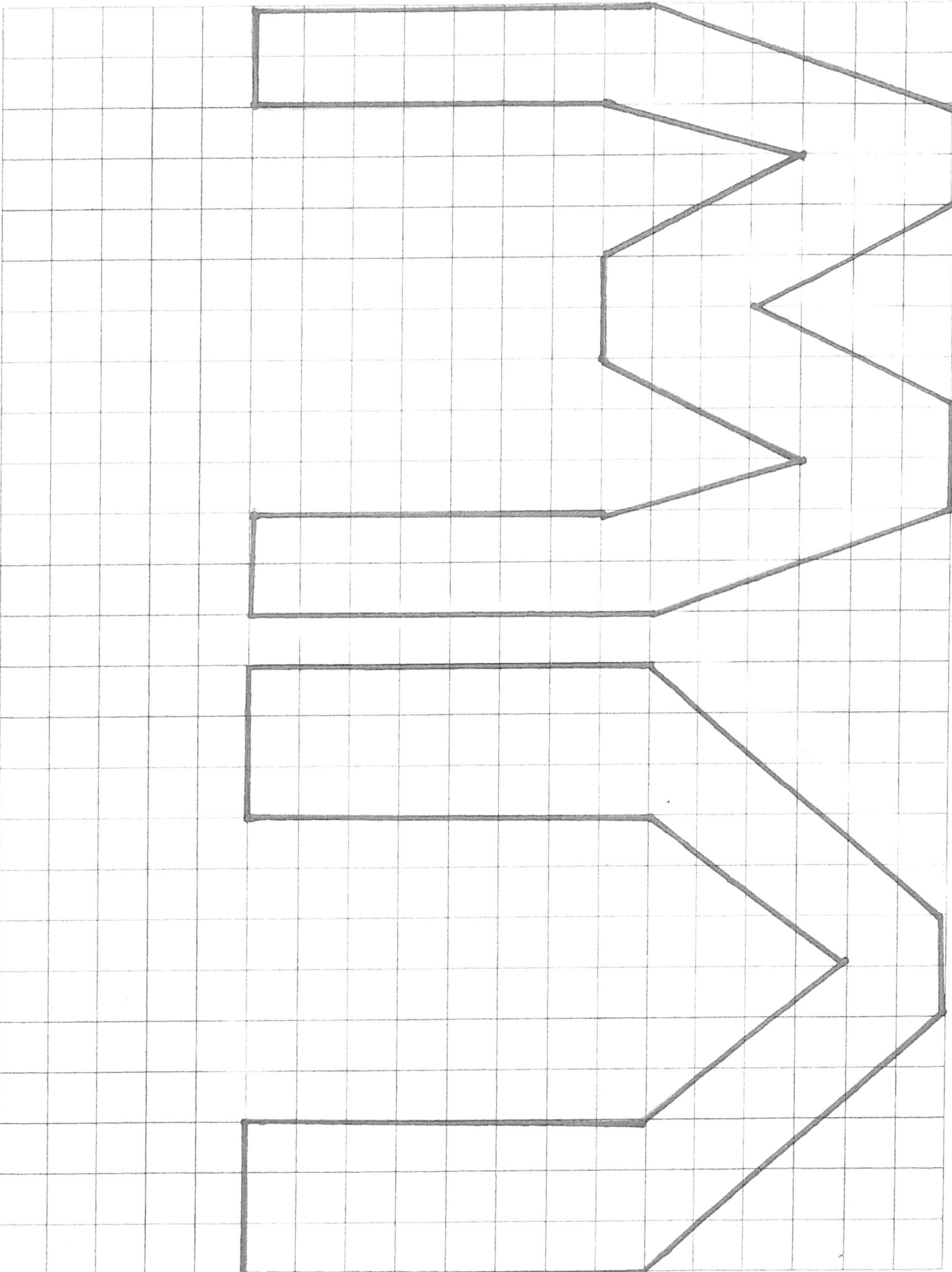

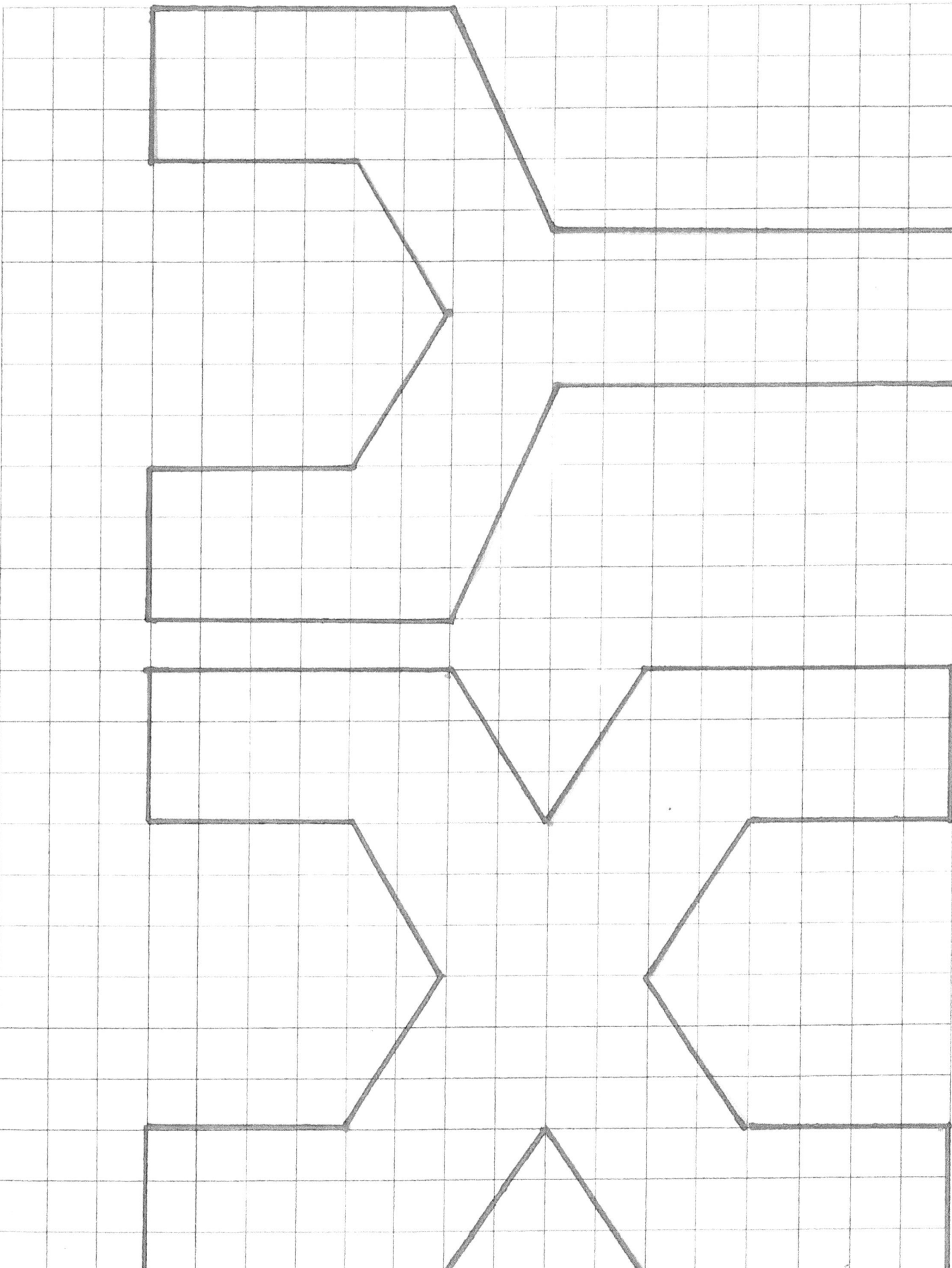

MATH... IT'S IN YOU!

This is a math vocabulary activity. My students love their name. They also love having their name displayed in the classroom. It gives them ownership to our room. Using my name....

T H O M A S – I can spell MATH with my last name.

K I M B E R L Y   E L L E N   T H O M A S   I can spell LINEAR with my full name.

Students have fun figuring out the math word(s) in everyone's name!

I have students write their name out on scratch paper and figure out what math word(s) they can spell using the letters of their name.

Then students will stencil their name on the drawing paper. Have them use a marker to color in the letters used for the math term a different color than the other letters not being used.

On the back of the worksheet, have students explain or give an example to prove they know the meaning of the vocabulary term they found.
Optional: Have students create a story about their word.

Throughout the year, as students are discovering new terms, they will find other math words in their name.

Actual Classroom Example:

MATERIALS NEEDED:

letter stencils (I use 1.5 – 2 inch)

drawing paper

markers: 2 different colors per student

MATH... It's in YOU!

Write Your First Name: _____

Write Your Middle Name: _____

Write Your Last Name: _____

Use a dictionary to help you, and write all the math words you can find using the letters of your name.

_____   _____

_____   _____

_____   _____

_____   _____

_____   _____

_____   _____

Directions: Use stencils to trace onto drawing paper whichever first, middle and/or last name you found the math word you are using. Choose ONE word from above and color the <u>letters of the math word</u> all the same color. Then color the other unused letters a different color. Please only use two colors total.

For each word in your name, explain that you know the meaning by drawing a picture, give an example, or use it in a sentence on the back of this sheet.

MEflection symMEtry

Actual Classroom Example:

MATERIALS NEEDED:

any size letter stencils

drawing paper

markers

scissors

MEflection SymMEtry

Directions: Choose stencil letters, or draw the letters on your own, to spell out your first name. Trace them on a fold. You may use any style of letters you want.

Cut out your name. BE CAREFUL NOT CUT THROUGH THE FOLD!!!

In three of your letters, put a word that describes you. You must put it in both letters to have reflection symmetry. Be careful when writing your words. Use pencil first!

Decorate the other letters in your name so that they also have reflection symmetry. Be creative and have FUN.

"IF I HAD A FRACTION OF YOUR BRAIN....."

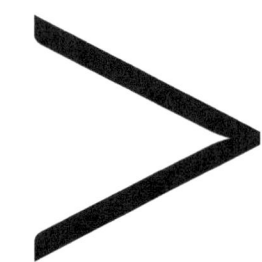

Fractions have never been a favorite among students. This project alleviates some fraction fears. To personalize this project, I trace or have students trace each other's silhouettes. I use an overhead or the Smartboard. You can use a generic silhouette or just an outline of a brain if you don't have time to draw the silhouettes.

Students divide their brain into 4 or more DIFFERENT fractions. The fractions they use must equal 1. I have students pair up to check if their fractions equal one whole. All the fractions must be in simplest form. When labeling their brain, the largest fraction must represent the largest part of the brain. They complete the activity sheet and then decorate their brain or silhouette to represent themselves.

Students love to see these displayed around the school. It gives them a sense of pride and ownership. As a teacher, it is great to learn more about my kids. You will enjoy reading about whose brain they would like to have. The kids enjoy guessing which silhouette belongs to their classmates. Modify according to grade level.

Actual Classroom Examples:

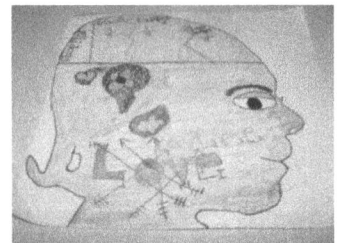

MATERIALS NEEDED:

Square tag board or paper, large enough for students' silhouettes. 1 piece for each student and 1 for the teacher. Kids love it when the teacher does the same work as them.

Overhead or smartboard projector

Markers, crayons, colored pencils

TIME:

Once every student has their silhouette, this takes about 80 minutes to complete.

"IF I Had a FRACTION of Your Brain...."

Name: _____

Directions: Divide up your brain into 4 or more pieces that reflect you. The fractions you use MUST equal 1 and MUST be different. Before you begin writing on your silhouette, use scratch paper to make sure your fractions equal 1. Have your partner critique your work. When labeling your brain, your fraction with the most value must represent the largest part of your brain. Your fraction with the least value must represent the smallest piece of your brain and so on. You may label your brain fractions with any appropriate labels that reflect YOU!! Be proud of yourself. Complete this activity sheet. Then, you may use crayons, colored pencils, and or markers to decorate your silhouette.

A. On the first line, write the fractions you used in order, from least to greatest. On the second line, prove that the fractions add up to 1 by adding them together:

_____

_____

B. Create problems about your brain so that the words **sum, difference, product, and quotient** must be applied to solve the problem. Have your partner solve each problem.

1. _____   2. _____

   _____   _____

   _____   _____

   Solution_____   Solution_____

3. _____   4. _____

   _____   _____

   _____   _____

   Solution_____   Solution_____

C. If you could have a fraction of anyone's brain, who would you choose and why?

_____

_____

_____

SHOP TIL YOU DROP SCAVENGER HUNT

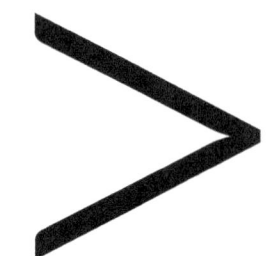

This project helps students understand how to apply tax and discounts to items they purchase. It also gives them the real life experience of writing a check. Fill teacher's name in for problem 7. It is always fun to see what they pick out for you. Use your state's sales tax rate.

Actual Classroom Example:

MATERIALS NEEDED:

variety of kid-friendly ads

scissors, glue or tape

construction paper

kid-friendly checks (online or templates given)

TIME:

This takes about two 80 minute periods to complete.

TEMPLATES:

Following the worksheet on the next page, a template for kid-friendly checks is provided as a resource to you.

Shop Til You Drop Scavenger Hunt

Name: _____

Ready for some mathlicious shopping?!! Look through the ads to find items that represent each situation. Cut out the item AND the PRICE!! Glue or tape the item to the construction paper. Please label it the appropriate number.

ALL SALES TAX is ____%. Write a check for each of the total costs. Tape or glue it to the construction paper next to the item.

1. Find TWO DIFFERENT items that cost the same. Compute the sum. Figure the sales tax and total cost.

Sum_____   Tax_____   Total Cost_____

2. Find 3 items that together cost less than $80.00. Compute the sum. Figure the sales tax and total cost.

Sum_____   Tax_____   Total Cost_____

3. Find two items that the difference in price is $10.00. Compute the sum. Figure the sales tax and total cost.

Sum_____   Tax_____   Total Cost_____

Shop Til You Drop Scavenger Hunt

(cont'd)

Name: _____

4. Find one item that costs between $40.00 and $60.00. Take 20% off.

Discount_____   Sale Price_____   Tax_____   Total Cost_____

5. Find two items that the difference in price is $20.00. Purchase the higher priced item. Take 25% off.

Discount_____   Sale Price_____   Tax_____   Total Cost_____

6. Find ONE item that you would like to have. Take 30% off.

Discount_____   Sale Price_____   Tax_____   Total Cost_____

7. Find ONE item that you would buy your teacher, _____. Take 15% off.

Discount_____   Sale Price_____   Tax_____   Total Cost_____

123 Mathlicious Ave.   Date_____

Pay to the Order of_____ $_____

_____Dollars

Bands Bank

For_____   _____

---

123 Mathlicious Ave.   Date_____

Pay to the Order of_____ $_____

_____Dollars

Bands Bank

For_____   _____

---

123 Mathlicious Ave.   Date_____

Pay to the Order of_____ $_____

_____Dollars

Bands Bank

For_____   _____

---

123 Mathlicious Ave.   Date_____

Pay to the Order of_____ $_____

_____Dollars

Bands Bank

For_____   _____

CIRCLIZE YOURSELF

This project allows students to apply their knowledge of radii, diameters, and chords. They will also apply the use of formulas for area and circumference.

Notice: Communicate that every line students draw must be a radius, diameter, or chord. Then they color what they need to spell their name.

Actual Classroom Examples:

THOMAS Shown first with the math elements sketched, then the letters colored in.

TONIO

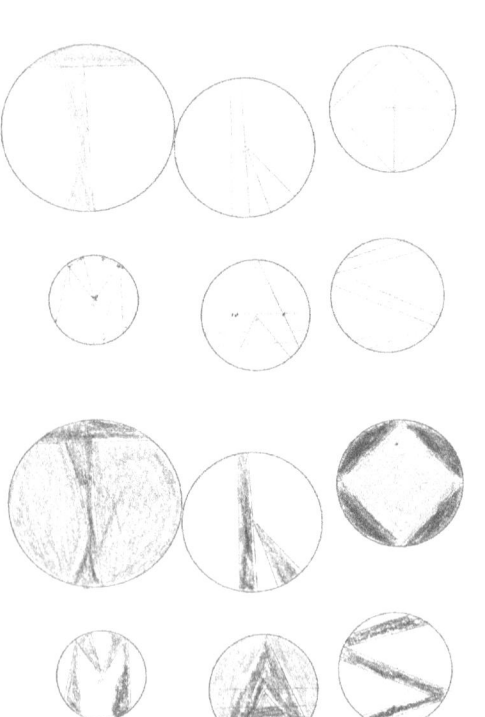

ARDALIONE Notice how he did a mathtastic job coloring in only what he needed to spell his name!!

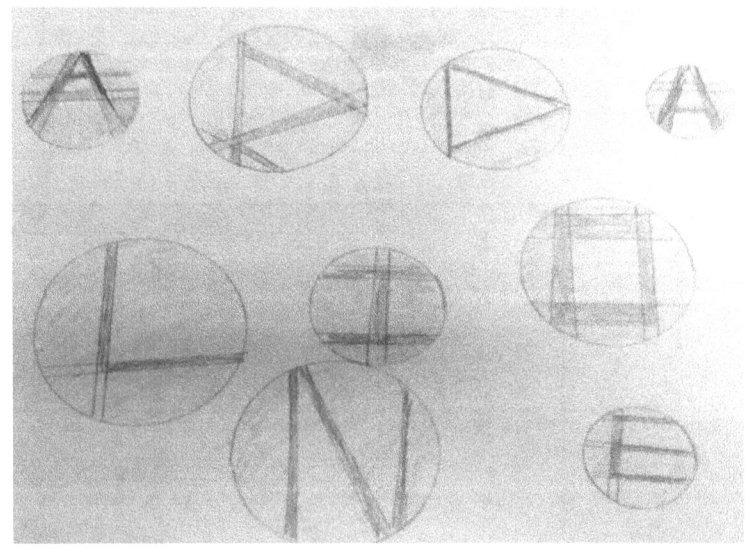

MATERIALS NEEDED:

safety compass / ruler

crayons, markers, or colored pencils

1 piece of copy paper per student

TIME:

This takes about 80 minutes to complete.

Circlize Yourself

Name: _____

Directions: Using the safety compass, construct different size circles. No circles can be congruent. You need as many circles as the letters in your name. Measure the length of each radii to the nearest tenth of a centimeter. Record the lengths on this sheet. Apply the formulas to calculate the area and circumference for each circle. If necessary, please round to the nearest hundredth. After you have drawn all of your circles and measured the radii, draw diameters, chords and/or radii in each of your circles to construct the letters of <u>your name</u>. Color the letters of your name the same color. Use any other colors for the area of the circle not used by the letters.

Letter	Radius	Area	Circumference
_____	_____	_____	_____
_____	_____	_____	_____
_____	_____	_____	_____
_____	_____	_____	_____
_____	_____	_____	_____
_____	_____	_____	_____
_____	_____	_____	_____
_____	_____	_____	_____
_____	_____	_____	_____
_____	_____	_____	_____
_____	_____	_____	_____
_____	_____	_____	_____
_____	_____	_____	_____
_____	_____	_____	_____
_____	_____	_____	_____

Be proud of your work!!

TANTALIZING TRANSLATIONS

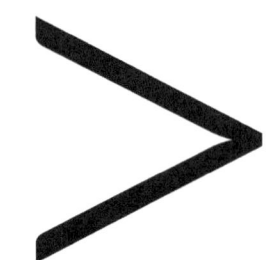

Students will apply their knowledge of how to translate a figure. They will create their own rules.

MATERIALS NEEDED:

1 piece of 4 quadrant graph paper per student

crayons, markers, colored pencils

TIME:

This takes about 80 minutes to complete.

Tantalizing Translations

Name: _____

Directions:
Plot **three** points in quadrant 1 to create a triangle.
Plot **four** points in quadrant 2 to create a quadrilateral.
Plot **five** points in quadrant 3 to create a pentagon.
Plot **six** points in quadrant 4 to create a hexagon.
Create your own rules for translating each figure. Write the coordinates for each image.

Quadrant 1 Triangle Coordinates _____ _____ _____

      Translation Rule   _____

      Image Coordinates _____ _____ _____

Quadrant 2 Quadrilateral Coordinates _____ _____ _____ _____

      Translation Rule   _____

      Image Coordinates _____ _____ _____ _____

Quadrant 3 Pentagon Coordinates _____ _____ _____ _____ _____

      Translation Rule   _____

      Image Coordinates _____ _____ _____ _____ _____

Quadrant 4 Hexagon Coordinates _____ _____ _____ _____ _____ _____

      Translation Rule   _____

      Image Coordinates _____ _____ _____ _____ _____ _____

Have a   **terrrrific**   **translated**   **day!!!**

ROCKIN' REFLECTIONS

Students will apply their knowledge of how to reflect figures over the x and y axis.

Actual Classroom Example:

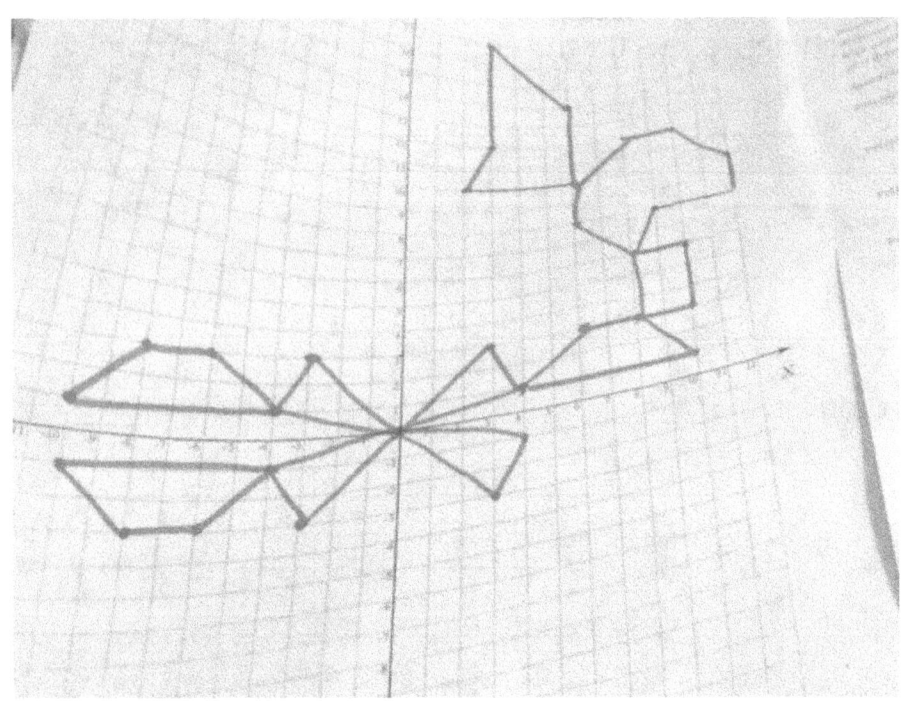

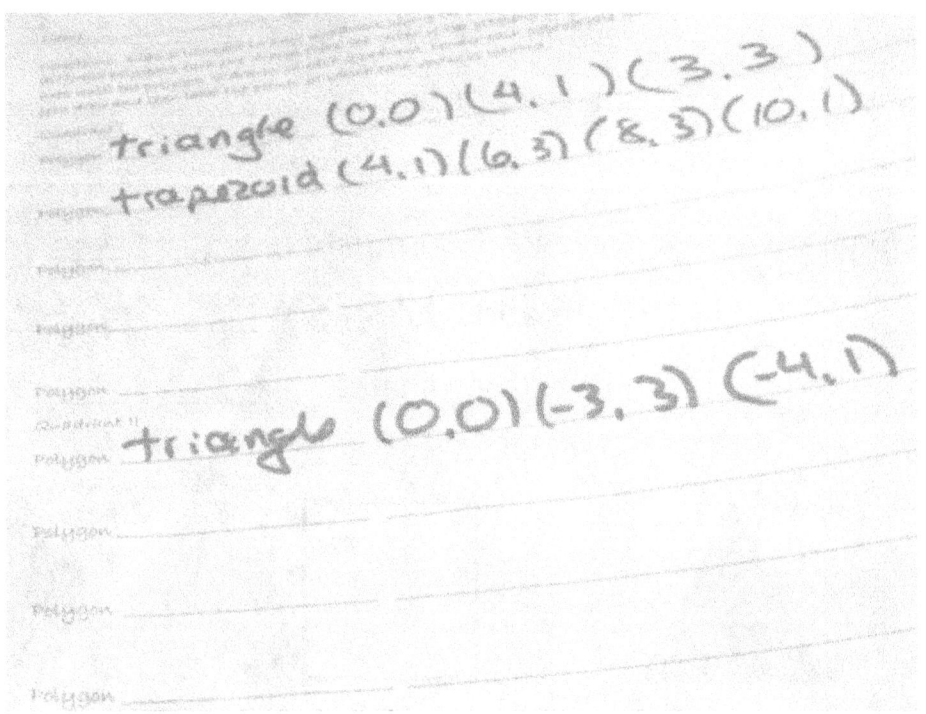

MATERIALS NEEDED:

1 piece of 4 quadrant graph paper

markers, colored pencils, crayons

TIME:

This takes about 80 minutes to complete.

R^2 = Rockin' Reflections

Name: _____

Directions: Draw a triangle in quadrant 1 using the origin (0,0) as one vertex. Write the name of the triangle by polygon under Quadrant 1. List the coordinates of each vertex. Reflect the triangle over the y axis. Now it is in quadrant 2. Write the name of the triangle by polygon under Quadrant 2. List the coordinates. You will reflect the triangle so that is located in each quadrant. Draw four other different polygons, each one drawn from a vertex of the previous one. Reflect them all so they are located in each quadrant. List the names of each polygon and the coordinates of each vertices under the appropriate quadrant. Color and have fun!!

Quadrant I   **Coordinates**

Polygon _____   _____

Polygon _____   _____

Polygon _____   _____

Polygon _____   _____

Polygon _____   _____

Quadrant II   **Coordinates**

Polygon _____   _____

Polygon _____   _____

Polygon _____   _____

Polygon _____   _____

Polygon _____   _____

Quadrant III   **Coordinates**

Polygon _____   _____

Polygon _____   _____

Polygon _____   _____

Polygon _____   _____

Polygon _____   _____

Quadrant IV   **Coordinates**

Polygon _____   _____

Polygon _____   _____

Polygon _____   _____

Polygon _____   _____

Polygon _____   _____

Have a day full of great reflections!!

TRANSINFORMATIONS

Students will apply their knowledge of transforming a figure on a coordinate plane.

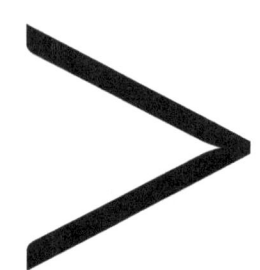

MATERIALS NEEDED:

4 quadrant graph paper (enlarged if possible, tabloid paper = 11"x17"): 1 for each student

markers, crayons, or colored pencils

TIME:

This takes about 80 minutes to complete.

TransINFORMations

Name: _____

1. Plot a square in quadrant 1 to represent your favorite theme park ride.

List the coordinates.

_____ _____ _____ _____

What is your favorite theme park ride?

_____

Create a dilation by using a scale factor of 2 (double) or 0.5 (half).

Plot the image (on the graph)

Write the new coordinates.

_____ _____ _____ _____

2. Plot a rectangle (non-square) in quadrant 2 to represent your favorite restaurant.

List the coordinates.

_____ _____ _____ _____

What is your favorite restaurant?

_____

Reflect the restaurant over the x axis.

Coordinates _____ _____ _____ _____

Reflect the restaurant over the y axis.

Coordinates _____ _____ _____ _____

3. Plot a trapezoid in quadrant 3 to represent your favorite store.

List the coordinates.

_____ _____ _____ _____

What is your favorite store?

_____

Translate 5 up and 3 right.

Coordinates _____ _____ _____ _____

4. Plot a triangle in quadrant 4 to represent a movie theatre.

List the coordinates.

_____ _____ _____

What is your favorite movie?

_____

Rotate the triangle 90° clockwise about the origin.

New Coordinates _____ _____ _____

Rotate the triangle 90° about any vertex.

New Coordinates _____ _____ _____

SIMILAR WHO? SIMILAR YOU!

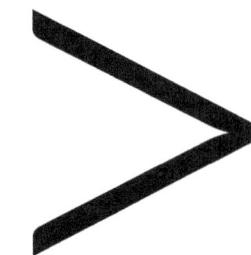

This project allows students to apply their knowledge of how to create similar figures. Students will create similar figures and record the measurements and scale factors. They will use the figures to spell out their name.

Actual Classroom Examples:

MRS. THOMAS

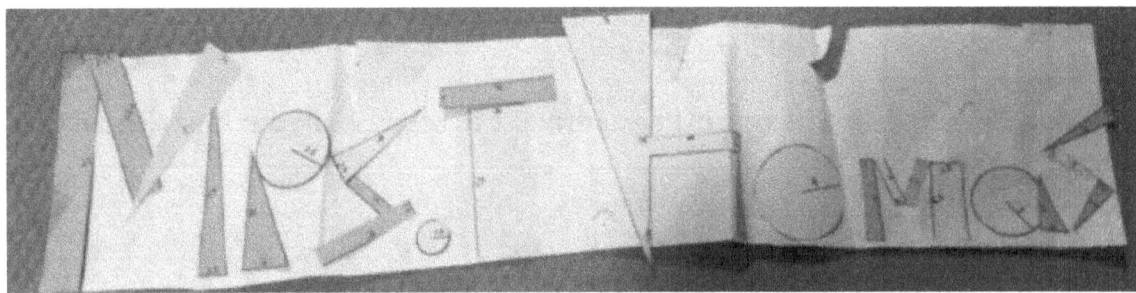

NIKYA

MARQUAY

MATERIALS NEEDED:

rulers, scissors, glue sticks, a variety of color paper

per student: 1 piece of bulletin board paper cut into strips about 1 foot long and then the width of the paper needed for each student's name

TIME:

This takes about three 80 minute periods to complete.

Similar WHO? Similar YOU!

Name: _____

Directions: Using a centimeter ruler, draw a shape. Write the dimensions on the shape. Use a scale factor to make the shape a smaller similar figure or a larger similar figure. Record the shape, original dimensions, scale factor and similar shape dimensions on this sheet for each set of similar figures you create. Please use at least 4 **different** scale factors, then you may repeat. NONE OF THE SHAPES CAN BE CONGRUENT.

After you have created several shapes and recorded each one, cut them out and use them to create your first name. Use a glue stick to glue it on the bulletin board paper.

Examples:

Shape	Original Dimensions	Scale Factor	Similar Dimensions
Triangle	b = 2.7cm h = 4.6cm	2	b = 5.4cm h = 9.2cm
Circle	r = 6cm	.5 or ½	r = 3cm
Rectangle	l = 4cm w = 5cm	3	l = 12cm w = 15cm

Shape	Original Dimensions	Scale Factor	Similar Dimensions
_____	_____	_____	_____
_____	_____	_____	_____
_____	_____	_____	_____
_____	_____	_____	_____
_____	_____	_____	_____
_____	_____	_____	_____
_____	_____	_____	_____
_____	_____	_____	_____
_____	_____	_____	_____
_____	_____	_____	_____
_____	_____	_____	_____
_____	_____	_____	_____

NAME YOUR GAME

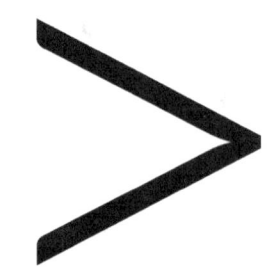

This is an AMATHAZING way for students to create a board game based on a variety of math standards. This is intended to be done in teams or groups. Ideally no more than 3 per group.

You can decide if you want focus on certain math concepts (to be learned) as a requirement of the game.

MATERIALS NEEDED:

scissors, markers, rulers, tag board, glue sticks, tape

1 poster board per group

possibly dice and spinner templates

computer access will allow students to type directions

TIME:

This takes about a week's worth of math classes

Name Your Game

Name(s): _____   _____   _____

Directions: Create a MATHlicious board game. Please write the directions for your game on this sheet of paper. Use the back if necessary. Construct places on your board for your game pieces to occupy. Give your game a name that would make people want to play it. You may use dice and/or create a spinner and/or create cards. Your group will present the game to the class. Everyone will get an opportunity to play all the MATHlicious games.

GAME NAME: _____

WRITE YOUR GAME'S RULES & DIRECTIONS:

MATHERPIECES

Students will create any object (piece of art) or their name by tracing numbers. They will complete an activity sheet according to the SUM they chose.

Modify according to students' needs.

Actual Classroom Examples:

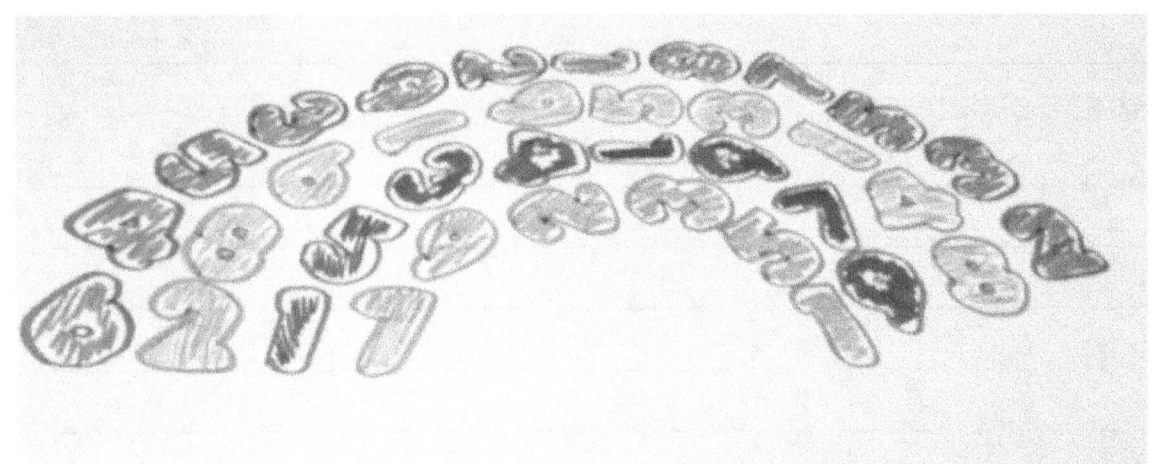

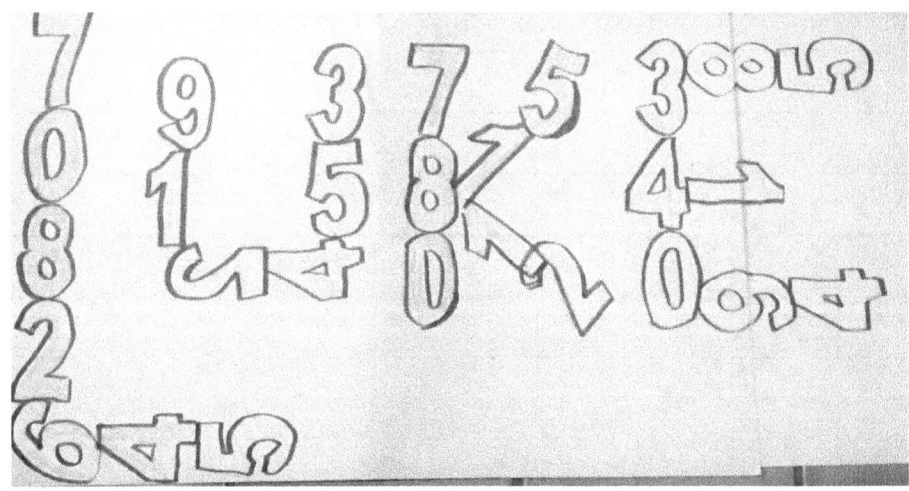

MATERIALS NEEDED:

1/2 sheet of poster board per student

different styles of number stencils

crayons, markers, colored pencils

TIME:

This takes about two 80 minute periods.

Matherpieces

Name: _____

Directions: Create a piece of art or your name by tracing number stencils. It can be anything: a robot, an animal, a fictional character, a flower, or anything you think of. Add up the numbers you use, and keep it a secret. You will present these to the class. The class will have to figure out your sum!

Use the sum of your Matherpiece to answer each question below.

1. Is your number prime or composite? _____ because _____
_____

2. Is your number even or odd? _____ because _____

3. Draw and label a **rectangle** so that it has the perimeter of your sum:

4. Draw and label a **triangle** so that it has the perimeter of your sum:

5. Draw and label a **rectangle** so that it has the area of your sum:

6. Draw and label a **triangle** so that it has the area of your sum:

7. Is the square root of your number a perfect square? _____ because _____
_____

8. Create an order of operations problem with at least 5 steps so that the answer is your sum:

ALWAYS TIME FOR MATH!!

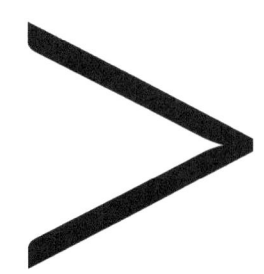

This activity allows students to use their prior knowledge in creating a MATH clock. Each number on the clock will be created using rules decided upon by the teacher or students. If you don't have a math clock, just cover up the numbers on the school clock with examples from the clocks the students create.

This can be adapted to any grade level!!

SAMPLE RULES FOR EACH HOUR ON THE CLOCK
(as seen in the example project to the right):

1:00	compute subtraction
2:00	compute a square root
3:00	compute two prime numbers
4:00	compute exponents
5:00	compute multiplication and division
6:00	compute addition and division
7:00	compute a ratio
8:00	apply properties of polygon(s)
9:00	compute an equation
10:00	compute three different numbers using two different operations
11:00	compute distributive property
12:00	think of one on your own that hasn't been used: _____

Actual Classroom Example:

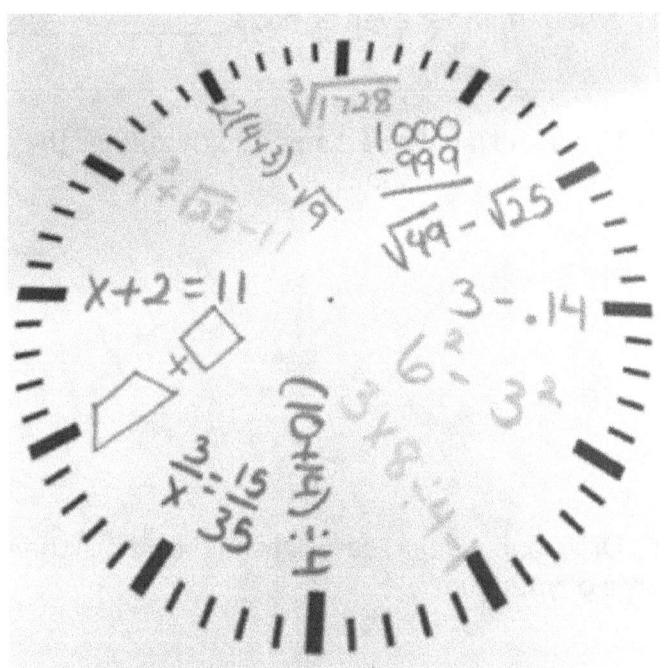

Always TIME for MATH!!

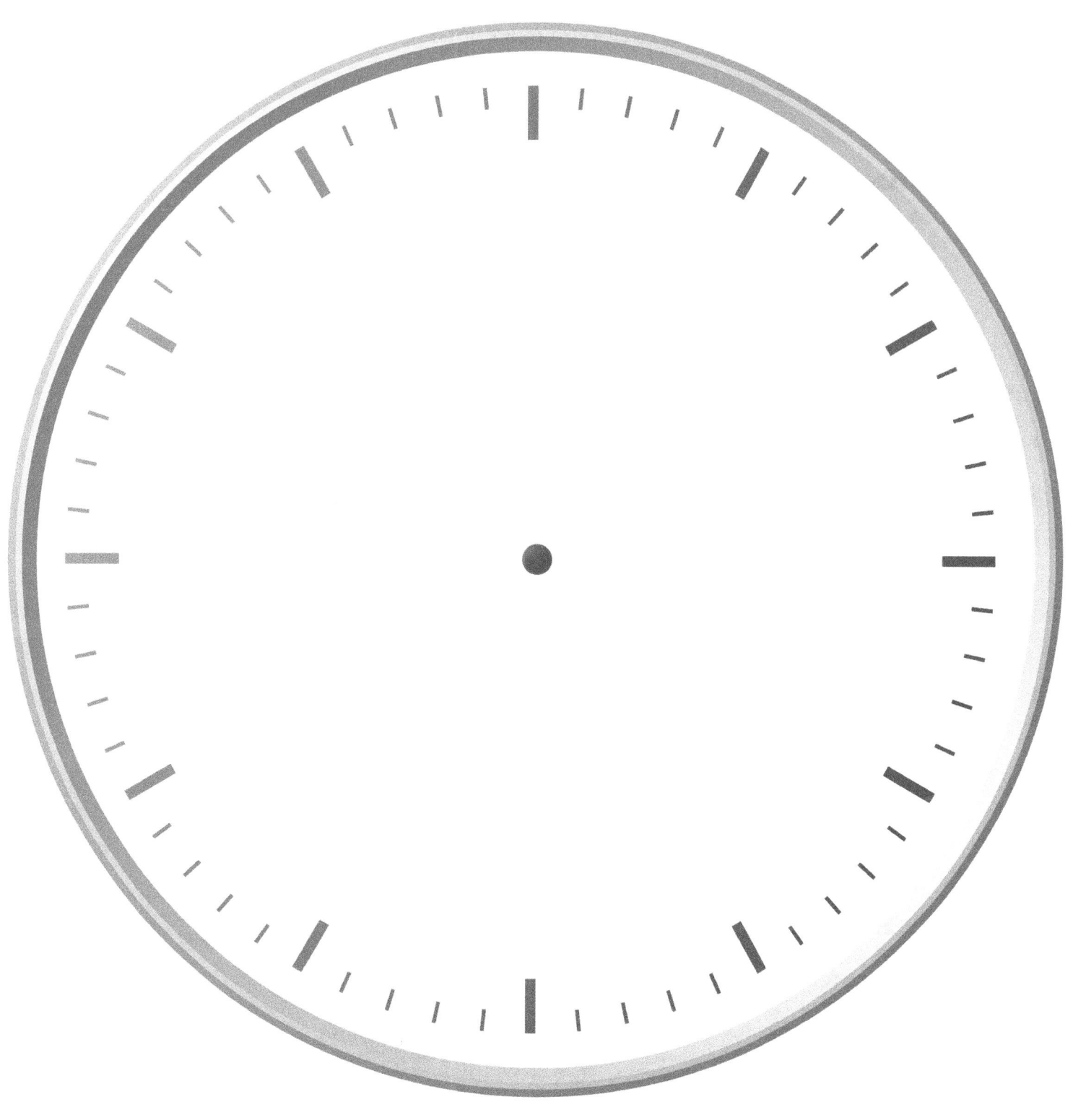

Name:

GEO IN THE HOOD

This project allows students to create their own neighborhood by applying knowledge of geometry.

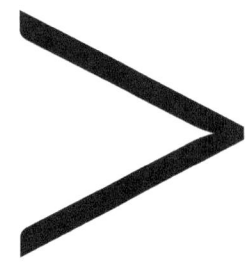

Actual Classroom Examples:

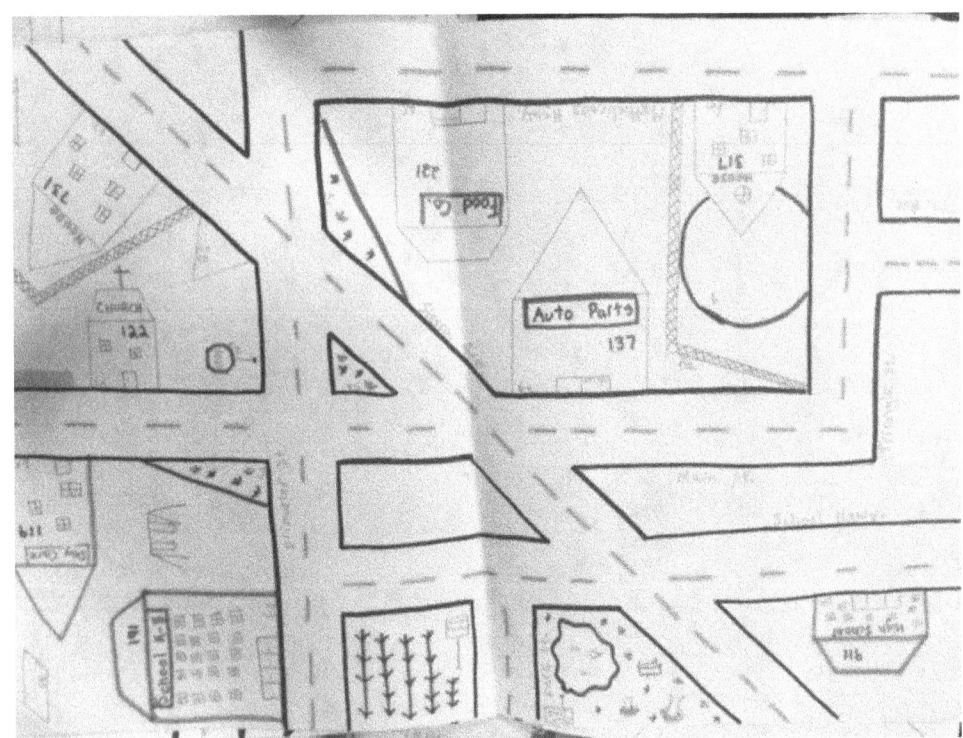

MATERIALS NEEDED:

large drawing paper for each student

ruler, safety compass

crayons, colored pencils or markers

activity sheet

TIME:

This takes about three 80 minute periods.

GEO in the Hood!

Name: _____

Draw and construct a version of your own Neighborhood of Geometry. Please use a ruler and safety compass. You may also use colored pencils, crayons, and/or markers. Follow the guidelines given below.

You are in charge of naming your neighborhood. You may have a theme and be creative. All of your streets and buildings must be named. Number each building a three-digit prime number. Create your streets using the width of the ruler.

Your neighborhood should include:
1. minimum of 8 streets
2. 3 streets must be parallel
3. 3 streets must intersect with each other
4. 2 streets must be perpendicular to each other

Label 5 – 7 on your neighborhood.
*Example: 5a should be **labeled** 5a on the right isosceles triangle you create.*

5. Create 4 parks in the shapes of triangles
    a. 1 right isosceles
    b. 1 acute equilateral
    c. 1 obtuse isosceles
    d. 1 right scalene
6. Create 8 buildings (houses, stores, library, restaurant…)
    4 in the shape of squares
    a. 1 with an area of 9 cm^2
    b. 1 with an area of 16 cm^2
    c. 1 with an area of 25 cm^2
    d. 1 with an area of 36 cm^2
    4 rectangle buildings
    e. 1 with an area of 12 cm^2
    f. 1 with an area of 20 cm^2
    g. 1 with an area of 24 cm^2
    h. 1 with an area of 10 cm^2
7. Create 2 sandboxes somewhere in your neighborhood
    c. 1 trapezoid
    d. 1 parallelogram that is not a square or rectangle

8. Create 1 circular swimming pool

9. Add 1 octagon stop sign

HAVE FUN!! BE PROUD OF YOUR WORK!!
Let your art skills and math knowledge SHINE!!

CHECKBOOK CHALLENGE

This teaches students how to balance a checkbook and budget their money. I ask local banks to donate checkbook covers and registers. Students live a month on minimum wage. (This is a 5 week project. I spend 10 -15 minutes on it each day.) Kids enjoy learning the adult way to budget money. We pretend everyone graduated high school and wants to move out and live on their own. I start everyone out with $600 and then a paycheck. Use whatever the minimum wage salary is in your state and figure 40 hours a week. Also be sure to take deductions out of checks. This is a great opportunity to explain taxes!! Every Friday, I have students deposit their paychecks by putting Deposit on the board and the amount they would have received for working 40 hours at a minimum-wage paying job. They don't actually get a check. I also put a gas bill on the board for an estimated amount they would have used in a week.

MATERIALS NEEDED:

set of checks (template provided, or lots of free different types are available online), register and cover for each student

create a purchase book by using blank paper stapled together OR fold construction paper sheets in half & distribute as needed

bucket bills (sample ideas provided after the Checkbook Challenge Due sheet in this book)

each day you should write a "bill" on the board.... such as: electricity, phone bill, cable, internet, car payment, insurance, water, etc.... anything "real life"

have lots of ads for students to look through or have them bring some to class

Checkbook Challenge Due sheet (next page)

I give all students 20 checks to begin the project (template follows the bucket bills). Please let them know to ask when they need more. They can number their own checks or you can put the numbers on them.

TIME:

This is a 5 week project. I spend 10 -15 minutes on it each day.

Checkbook Challenge

Name: _____   Due: _____

**FOR ALL THE DATES LISTED BELOW, WE WILL WORK TOGETHER IN CLASS.
BE PREPARED, and YOU WILL BE ABLE TO COMPLETE MOST OF THIS IN CLASS!**

Each day listed below, you will write a $65.00 check for groceries. You may bring in $15 worth of coupons. Your coupon amount will be deducted from the $65.00. You may not use a coupon for the same item twice on the same Monday. All coupons must be for something you would eat and not expired.

Monday, _____   Monday, _____   Monday, _____   Monday, _____

Due _____: Find a 2 bedroom apartment or house to rent. Cut out the ad and tape it in your purchase log. We will do this together in class. If the deposit is not listed, make it the same as the rent. You will add the rent and deposit together, then divide by 2. You can't use any free rent ads!!

   Items you need to budget for during this project:

Due _____: Cell phone – Cut out ad and tape it in your purchase log. Figure ____ % sales tax.

Due _____: Television – Cut out ad and tape it in your purchase log. Figure ____% sales tax.

Due _____: Furniture ONE piece (something you would sit on) – Cut out ad and tape it in your purchase log. Figure ____% sales tax.

Due _____: $60.00 check due for miscellaneous (paper and cleaning products, toiletries, soap, deodorant, shampoo, detergent, etc…) COUPONS may be used up to $20.00 worth. YOU DO NOT PUT ANY OF THESE ITEMS IN YOUR PURCHASE LOG. You just write a check to Walmart.

Due _____: Cookware (pots, frying pans, skillet, cookie sheet) – Cut out the ad and tape it in your purchase log. Figure ____% sales tax.

Due _____: Vacuum – Cut out ad and tape it in your purchase log. Figure ____% sales tax.

Sometime during this project, you must purchase a pair of shoes, a shirt, and shorts, pants or a skirt. Be sure to cut out a picture and price of what you are buying, and tape it in your purchase log. Figure_____ % sales tax. You may write one check for all three items.

Please use a calculator, but ALL sales tax calculation steps must be shown in your purchase log on the same page or across the page from where you taped the item you are buying.

You will have many other bills that will be on the board on a variety of days. You are responsible for any days you are absent. You may not write a check for something unless you have the money. If you don't have enough money to pay for something the day it is due, write the item on your "Could Not Afford" page in your purchase log. When you have enough money to pay for the item, write a check and cross it off. This will help keep you organized.

Every Thursday, you will pick a bill out of the bucket. All bucket bills MUST be taped to the same page! All bucket bills must be different.

REMEMBER, even if you can't afford an item you are supposed to purchase, you still MUST have an ad for the item cut out and taped in your purchase log with the sales tax calculations. Any item you could not afford must be written on the Could Not Afford page. If you budget wisely you will not have many items on this page.

Learn how to budget!!   Have Fun Being On Your Own!

CHECKBOOK CHALLENGE BUCKET BILLS

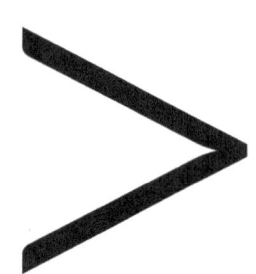

This is a supplement for the Checkbook Challenge.

Please adapt to the needs of your students (certain bills you can fill in amount). I cut these up, and put them in a bucket or container. They pick one each week. They also tape it in their purchase log:

Oops… You were not wearing your seatbelt. Pay City Circuit Clerk $___.00.

You were sick and went to the doctor. Write a check to Dr. Sniffles for $50.00.

It is time for an oil change. Write a check to Sears Automotive $26.97.

You rented 3 DVDs & forgot to return them. Write a check to Vertex Video $23.94 for a late charge!

You went out with friends. Withdraw $40.00.

Your Visa monthly payment is due. Write a check to Visa for $38.71.

You bought a new video game. Write a check to Game Stop for $59.53

You had a toothache. Write a check to Dr. Snaggletooth for $95.00.

You ordered pizza and wings. Write a check to Domino's for $16.88.

Oops, you got caught speeding. Write a check to City Circuit Clerk for $___.00.

You went over your data limit on your cell phone. Write a check to Verizon for $47.32.

Your Discover card payment is due. Write a check to Discover for $41.17

123 Mathlicious Ave.   Date_____

Pay to the Order of_____ $_____

_____Dollars

Bands Bank

For_____   _____

123 Mathlicious Ave.   Date_____

Pay to the Order of_____ $_____

_____Dollars

Bands Bank

For_____   _____

123 Mathlicious Ave.   Date_____

Pay to the Order of_____ $_____

_____Dollars

Bands Bank

For_____   _____

123 Mathlicious Ave.   Date_____

Pay to the Order of_____ $_____

_____Dollars

Bands Bank

For_____   _____

THE MATHLICIOUS BOOK CLUB

DIRECTIONS:

- Read a children's math book as a group (3-4 in a group)

- Create an assessment that the class could answer after your club reads them the book. The assessment must have 5 questions. Each person needs to create one question. Decide as a group what the other questions should be (to total 5). Make an answer key on a separate sheet of lined paper.

- As a club, choose 4 math terms from the book. Assign each person a term. Each person needs to:

    - prove they know the meaning of the word

    - create a fracordiddle (see page 18)

        Example: 2/3 of rap + 1/3 of diaper + 1/2 of dust = radius

    - create a camathouflage using between 16 and 20 letters (see page 13)

        Underline the letters of the word

PRESENTATION – Your club will present your project to the class!

- Present the camathouflage and fracordiddles

- Read the book to the class

- Pass out assessment, then grade them.

Book Club Form

This Is Due: _____

Name of Your Book Club: _____

Members: _____   _____

_____   _____

Assessment Questions

1. _____

2. _____

3. _____

4. _____

5. _____

Math Terms

1. _____   Proof you know the meaning (write on backside)

    Fracordiddle _____

    Camathouflage _____

2. _____   Proof you know the meaning (write on backside)

    Fracordiddle _____

    Camathouflage _____

3. _____   Proof you know the meaning (write on backside)

    Fracordiddle _____

    Camathouflage _____

4. _____   Proof you know the meaning (write on backside)

    Fracordiddle _____

    Camathouflage _____

MATHOLIDAY ACTIVITIES & PROJECTS

MATH'O'LANTERNS

Students will construct the math version of a Jack'o'Lantern. While participating in this project, students discover surface area and explore the properties of 3D figures. They also demonstrate the use of prior knowledge while computing area. Language arts is incorporated as students write a Limerick that describes their Math'o'Lantern.

Actual Classroom Example:

MATERIALS NEEDED:

1 cereal box per student

1 piece of orange bulletin board paper per student (big enough to cover the cereal box)

scissors, glue sticks, tape, ruler

nets to be cut out (search online)

lots of black construction paper for the nets to create facial features (plus any other colors to decorate and personalize)

TIME:

This takes about three 80 minute periods.

Math'o'Lanterns - Directions

- *Trace prism (cereal box) on orange bulletin board paper.*
- *I have kids on the floor in the classroom and out in the hallway.*
- *After tracing the net, have students cut it out.*

- *Measure and complete top portion of project worksheet (see next page).*
- *Glue paper to prism.*

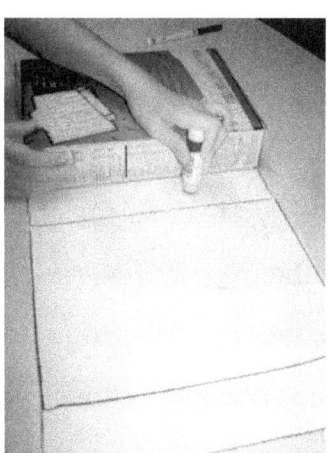
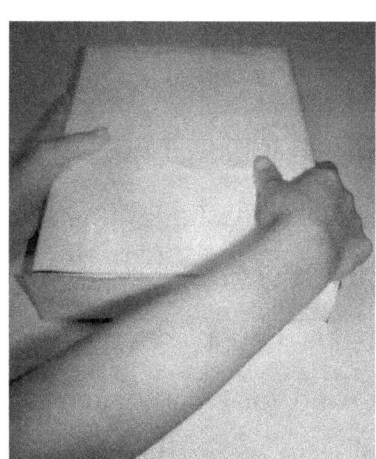

- *Students cut, measure, and fold nets to create facial features.*
- *Continue to complete project worksheet.*

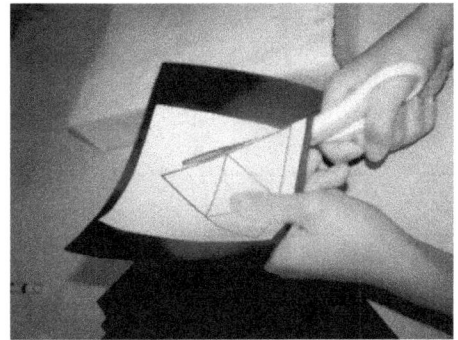
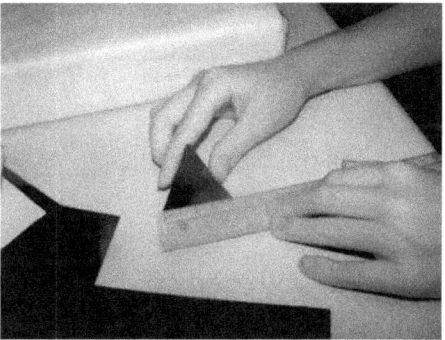

Math'o'Lanterns - Worksheet

Name: _____

1. Trace your prism to create the net. Using centimeters, measure the dimensions of your cereal box. How many different measurements will you have? _____

Record the measurements on your net, as well as this net:

    Length: _____

    Width: _____

    Height: _____

2. What is your surface area? _____

3. EXPLAIN how you computed surface area.

_____

_____

_____

4. Try to create a formula for finding surface area of any rectangular prism: _____

5. How many of each shape are congruent on your prism (cereal box)? _____

Create Your Math'o'Lantern Face...

Please use at least one cube and one pyramid. All facial features must be 3D.

6. Record the side length of your cube: _____

7. What is the surface area of the cube? _____

8. How did you compute the surface area of the cube? _____

_____

9. Record the base and height of the triangular face of the pyramid: Base _____ Height _____

10. If your pyramid has a square for its base, record the side length: _____

11. What is the area of the triangle? _____

12. What is the area of the square? _____

13. What is the surface area of the pyramid? _____

14. How did you compute the surface area of the pyramid? _____

_____

15. Write the names of all the other figures you used on your Math'o'Lantern: _____

_____

MATH'O'LANTERN LIMERICK

Name: _____

Example of a Limerick:

(A) There once was a math teacher with swag
(A) Who liked throwing things in a bag
(B) She plays a math game
(B) So class isn't lame
(A) And her students don't need to nag

Rough Draft of YOUR Limerick:

There once was a math'o'lantern named _____
_____
_____
_____
_____

My Math'o'Lantern Limerick!

_____

_____

_____

_____

_____

HAPPY MATHO'WEEN

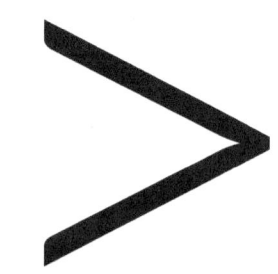

This activity allows students to create a math trick or puzzle. Students present these to the class on MathO'ween (Halloween party day). I assign this at least one week prior to the MathO'ween celebration. These can range from easy to very challenging. Modify according to grade level.

For the MathO'ween celebration, whoever figures the math trick gets a treat!

Actual Classroom Examples of Math Puzzles:

Ten ghosts were at a cemetery. Each of them brought candy to share. The ghosts were numbered 1 through 10:

- Each ghost with a prime number ate 5 Starbursts
- Each ghost with a number divisible by 3 ate 3 Snickers
- Each ghost with a perfect square number ate 10 Lifesavers

How many pieces of candy were eaten? _____

What two prime numbers have a sum of 30?
_____

A ghost named Geo went trick or treating. At the first house, he received $\sqrt{49}$ treats. At the second house, he received the smallest prime number of tricks. At the third house, he received the tenth composite number of treats. At the fourth house, he received 4^2 tricks. At the last house, he received today's date number of treats. What is the difference between Geo's treats and tricks? _____

What is the next number in this pattern?:

7

15

31

63

_____

TIME:

Students present these to the class on MathO'ween (Halloween party day). I assign this at least one week prior to the MathO'ween celebration.

 # Happy MathO'ween!

Name: _____

This is due on or before: _____

CREATE two or more grade appropriate MATH tricks.

Be prepared to present it to the class. Make sure you are able to explain how to get the solution. Show the tricks on this sheet. Have the answers on the back.

It could be a pattern problem, story problem, find the missing number, or any number trick (puzzle). Be TRICKY and have fun!!!

maTHANKFULness 2D VERSION

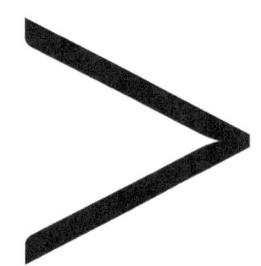

There are two versions of this project. This first version has students create a 2-dimensional turkey using a circle graph as the body. Students will construct the graph by using a protractor. Students fill out a project sheet while computing the percent, fraction, and degree of each sector. Modify according to the needs of your students.

See next page for the second version (3D turkey).

Actual Classroom Examples:

First, I have the students complete the top half of the activity sheet. I always have to remind my students that the percents need to equal 100, degrees = 360, and fraction = 1. I have them double check before they begin constructing their graph. They should turn the worksheet in with their turkey. They put the circle on top of the brown construction paper and cut it out. Then they use a protractor to construct each sector. If some of your students will struggle creating different sectors, this is a great project to differentiate the directions according to the needs of your students. The students will label each sector the five things they are thankful for: percent, degree, and fraction. They will cut out their own shape of feathers and write on them what they are thankful for. I allow them to personalize their turkey with other decorations that they want.

These look great posted around the classroom or school!

MATERIALS NEEDED:

1 piece of brown construction paper for each student (big enough for the circles)

several quarter sheets of different colored construction paper for the feathers

protractors, scissors, tape, glue, markers

TIME:

2D version takes about 80 minutes

MaTHANKFULness... Gobble It Up!! 2D

Name: _____   (also put your name on the back of your turkey)

Complete this sheet **before** you begin your circle graph. Create a circle graph (turkey) with 5 DIFFERENT sectors. Label each sector with its percent, fraction, degree, AND something you're THANKFUL for. ONLY ONE PERCENT CAN END IN ZERO. Make feathers to put on your turkey. You may label the feathers with the things you are thankful for. Draw or create a face, and make turkey legs too! Use a marker or pen to label your circle graph. This makes it easier to see.

BE CREATIVE AND GOBBLE UP ALL THIS GOOOOD MATH

Five things I am thankful for...	Percent	Degree	Fraction
_____	_____	_____	_____
_____	_____	_____	_____
_____	_____	_____	_____
_____	_____	_____	_____
_____	_____	_____	_____

Put this circle on your brown construction paper, and cut it out.

•

Push down with a pen or pencil on the center dot, so it will show through to the center of the brown paper.

MATHANKFULNESS 3D VERSION

There are two versions of this project. This second version has students create a 3-dimensional turkey using two circle graphs with a cylinder in-between as the body. Students will construct the graph by using a protractor. Students fill out a project sheet while computing the percent, fraction, and degree of each sector. Modify according to the needs of your students.

See previous page for the first version (2D turkey) AND the majority of the general project instructions.

What makes this version 3-dimensional is that two circles are cut out (front and back) & a tag board cylinder is added to give depth to the turkey.

The feathers are also made into cones, instead of flat feathers.

Actual Classroom Example:

MATERIALS NEEDED:

1 piece of brown construction paper for each student (big enough for the circles)

several quarter sheets of different colored construction paper for the feathers

protractors, scissors, tape, glue, markers

tag board (this is the only additional material needed for 3D version)

TIME:

3D version takes about two 80 minute class periods.

MaTHANKFULness... Gobble It Up!! 3D

Name: _____ (also put your name on the back of your turkey)

Complete this sheet **before** you begin your circle graph. Create a circle graph (turkey) with 5 DIFFERENT sectors. Label each sector with its percent, fraction, degree, AND something you're THANKFUL for. ONLY ONE PERCENT CAN END IN ZERO. Use tag board to create a cylinder and use the turkey body as one of the circular bases. Create cones for the feathers to put on your turkey. You may label the feathers with the things you are thankful for. Draw or create a face, and make turkey legs too! Use a marker or pen to label your circle graph. This makes it easier to see.

BE CREATIVE AND GOBBLE UP ALL THIS GOOOOD MATH

Five things I am thankful for...	Percent	Degree	Fraction
_____	_____	_____	_____
_____	_____	_____	_____
_____	_____	_____	_____
_____	_____	_____	_____
_____	_____	_____	_____

Calculate your cylinder's:

Radius _____

Height _____

Volume _____

Calculate one cone's:

Radius _____

Height _____

Volume _____

Put this circle on your brown construction paper and one piece of tag board, trace it, then cut it out. (To get the height for the 3D effect, cut a rectangular piece of tag board that equals the circumference of the circle)

Push down with a pen or pencil on the center dot, so it will show through to the center of the brown paper.

GOBBLING GRAPHS

Students will apply knowledge of graphing equations.

Actual Classroom Example:

MATERIALS NEEDED:

scissors

graph paper with circle on it

activity sheet

glue sticks

different color paper for feathers

TIME:

This takes about 80 minutes.

Gobbling Graphs

Name: _____ (also put your name on the back of your turkey)

Directions:
- Look at your graph paper to figure out what equations will fit.
- Write two equations with positive slopes.
- Write two equations with negative slopes.
- Graph each equation.
- Write two coordinates (solutions) of the equation that are on the graph.

	Equation	Slope	Y-Intercept	Coordinates
1.	_____	____	____	____
2.	_____	____	____	____
3.	_____	____	____	____
4.	_____	____	____	____

5. Write an equation parallel to equation 1. _____

Graph it please.

What makes it parallel?_____

Write two coordinates on the graph: _____ _____

6. Write an equation perpendicular to equation 2.

Graph it please.

What makes it perpendicular?_____

Write two coordinates on the graph: _____ _____

Cut out your graphs. Glue it to the brown circle. Cut out the brown circle. Cut out six feathers. Write equations 1 – 6 on each feather. Be creative and decorate the feathers. Glue the feathers to create a turkey. Create a head and feet for your turkey!

GEOBREAD HOUSES

This is a favorite project among all my students. This project takes two or three days, and the students take them home on the day after they are completed. You will need tables to store the houses overnight. I have done this project in the science lab. This can be adapted to most grade levels. I pass out the supply sheet before Thanksgiving break. I supply extras of everything. I also supply tape.

Teacher Instructions:
Day 1

Students tape the Kleenex box to the foil covered card board. Then they use the graham crackers to construct the walls and part of the roof. The frosting is put into the baggies and then cut a corner for piping. Use the frosting as the glue. Put two whole graham crackers up on each of the four faces of the Kleenex box. Use four whole graham crackers for part of the roof.

End of Day 1 -->

MATERIALS NEEDED:

Listed on the supply sheet you give to your students (see next page)

TIME:

Pass out supply sheet before Thanksgiving break. This project takes two or three days.

GEOBREAD HOUSES (CONT'D)

Teacher Instructions:
Day 2

The students finish the roof. Have students cut off two right triangles on the end of two whole graham crackers to create a trapezoid. Use the right triangles to fill in the roof on top of each trapezoid. Put the longer of the two legs together. The hypotenuse should meet the edge of the roof.

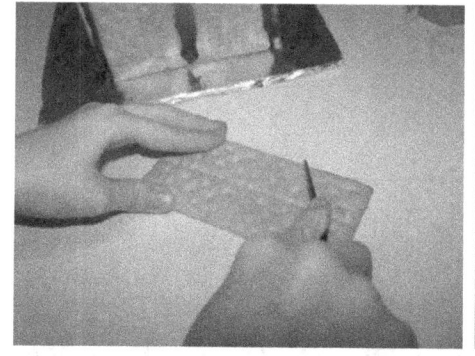
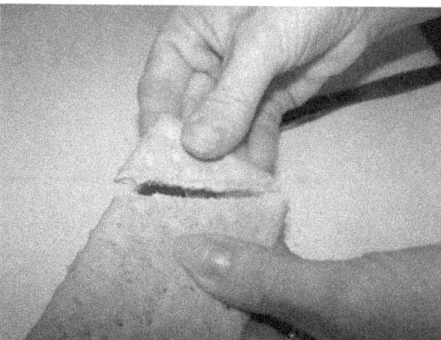

End of Day 2 -->

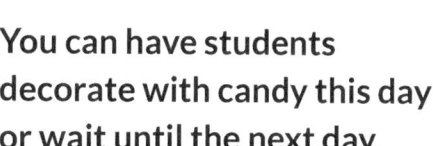

You can have students decorate with candy this day or wait until the next day.

Day 3

The students decorate their home with candy. -->
You can differentiate this day by having students decorate by creating different kinds of angles and lines.

Student should turn in their completed activity sheets from day 1-3 today.

GEObread Houses - Supply List

Name: _____

Directions:

Please have all of these supplies on or before_____.

Please bring all the supplies in at one time in a plastic grocery sack. Put your name on **everything**.

Here is a list of supplies you need to bring:

- 1 <u>SQUARE</u> kleenex box
- 1 unopened box of RECTANGULAR graham crackers
- 1 unopened plastic cylinder container of frosting (any flavor) NO TUBES
- 1 piece of about 10 inch by 10 inch cardboard or "pizza circle" covered in foil
- 3 sandwich bags that can be sealed
- 3 plastic spoons
- 1 plastic knife
- 1 large bag of CANDY. I am bringing: _____

    Everyone will bring one large bag of candy to share with the class. All the candy will be divided up among all the students so everyone will end up with their own mixed bag. Here are some suggestions, but you may bring what want:

Gum Drops	Gum Balls	Lemon Drops
Skittles	Caramel Cubes	Sour Balls
Hershey Kisses	Pull-n-Peel Twizzlers	Mike & Ikes
Starbursts	M&Ms	Marshmallows
Bugles	Tootsie Rolls	Mini Candy Canes
Jolly Ranchers	Dots	

This list is all the individual items you need, but you may bring as much of each item as you would like.

GEObread Houses - Activity Sheet

Name: _____

PART ONE

1. Measure to the nearest tenth of a centimeter the length and width of one graham cracker.

   Length _____ Width _____ Area _____ Perimeter _____

2. How many graham crackers did you use today? _____

3. So far, what is the surface area of your home? _____

4. List ALL possible shapes that classify your graham cracker:
_____

5. Create a story problem about the graham cracker:

_____

_____

_____

6. 2/5 of the students are using chocolate frosting. 1/4 of the students are using strawberry frosting. The rest are using vanilla. What percent are using vanilla frosting? _____

7. What did you enjoy about day one? _____

_____

PART TWO

1. Trace your trapezoid on this sheet. To the nearest cm, measure the bases and height. Label the measurements on your drawing.

Area _____

2. Trace your triangle. To the nearest cm, measure the base and height. Label the measurements on your drawing.

Area _____

GEObread Houses - Activity Sheet (cont'd)

Name: _____

3. What is the total surface area of your roof? _____
   Don't forget about the rectangles that make up part of your roof.

4. What did you enjoy about day two?_____
_____

PART THREE

List four different types of candy you used to decorate your house AND estimate the percent that each candy makes up on your house. Your percent total should equal 100!!! Create a circle graph that represents your house. All percent must be different. Label the graph with candy type, percent, degree, and fraction.

Candy	Percent
_____	_____
_____	_____
_____	_____
_____	_____
TOTAL	_____

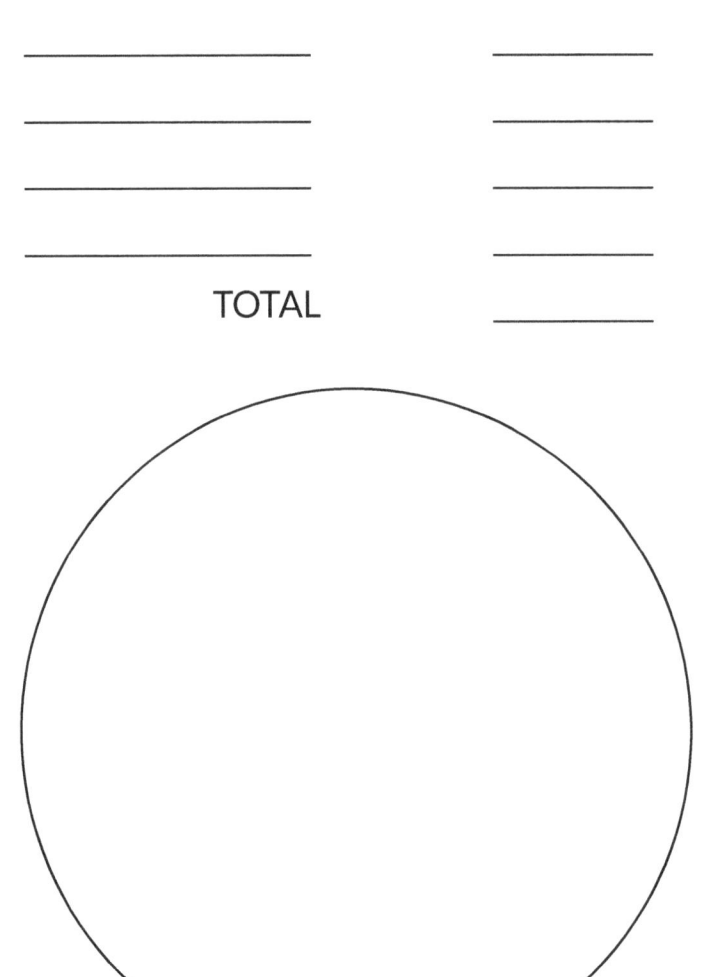

GEObread Houses
EXTENSION CHALLENGE ACTIVITY

Architect: _____

DESIGNING YOUR FLOORPLAN:

Trace the base of your Kleenex box on centimeter grid paper. Divide your house up into 4 to 7 rooms. Draw top view pictures to reveal each room. Find the area of each room using the scale 1cm = 2m. Record each room with its dimensions:

ROOM	Dimensions in cm	Dimensions in m	Area in meters
_____	_____	_____	_____
_____	_____	_____	_____
_____	_____	_____	_____
_____	_____	_____	_____
_____	_____	_____	_____
_____	_____	_____	_____
_____	_____	_____	_____

Example of an architect's floorplan...

$\sqrt{144}$ DAYS OF CHRISTMATH

I begin on December 1. Students create a problem applying the order of operations to equal the day you are on. Give each day the guidelines below that must be used to create that day's equation. After creating a problem that fits the description for the day, have students write something they think you would give them. On the final day, I sing this with my students to the tune of 12 Days of Christmas. Feel free to run with this & modify each daily guideline to fit the needs of your students. It is so mathtastic to have students share how they used each number to equal that day.

Daily Guidelines (starting place):

Day 1: Use a square root, 3 different numbers, and different operations to equal the day!!

Day 2: Use three different integers... Square one integer, use division and subtraction.

Day 3: Use two different numbers in parenthesis, then any two operations with any two numbers.

Day 4: Use four different numbers, one must have two digits.

Day 5: Use 10, 7, 6 at least once. Use any operations. You may use these more than once to equal 5.

Day 6: Use 21, 5, 14, 4 at least once. Use any operations.

Day 7: Use any three prime numbers and any operations to equal 7.

Day 8: Use two even and two odd numbers. Use a square root.

Day 9: Use at least 3 single-digit numbers and any operations.

Day 10: Use three different numbers with two different operations. Use an exponent.

Day 11: Use four different digits and at least one set of parentheses. Use any operations.

Day 12: Use the first four prime numbers (2, 3, 5, 7) with any operation.

Teacher Example:

On the $\sqrt{100} \div 5 - 1$ day of ChristMATH my math teacher gave to me... *math muscles for me to gain knowledge*.

On the $6^2 \div 9 - 2$ day of ChristMATH my math teacher gave to me... *two ways to solve for x*.

On the $4(2 + 3) - 17$ day of ChristMATH my math teacher gave to me... *three area formulas*.

On the $24 \div 8 + 4 - 3$ day of ChristMATH my math teacher gave to me... *four central tendencies*.

On the $(8 + 10) - (7 + 6)$ day of ChristMATH my math teacher gave to me... *five golden ratios*.

On the $21 - 14 - 5 + 4$ day of ChristMATH my math teacher gave to me... *six step equation*.

On the $5 \times 2 - 3$ day of ChristMATH my math teacher gave to me... *seven square roots*.

On the $\sqrt{121} + 8 \div 2 - 7$ day of ChristMATH my math teacher gave to me... *eight exponents*.

On the $8 + 4^2 - 5 \times 3$ day of ChristMATH my math teacher gave to me... *nine negative slopes*.

On the $8^2 - 60 + 6$ day of ChristMATH my math teacher gave to me... *ten tetrahedrons*.

On the $(2 \times 3) + (5^2 - 20)$ day of ChristMATH my math teacher gave to me... *eleven inequalities*.

On the $(7 + 5) \div (3 - 2)$ day of ChristMATH my math teacher gave to me... ***TWELVE HYPOTENUSE HUGS!!!***

TIME:

This takes about 10 minutes each day for the 12 days.

$\sqrt{144}$ Days of ChristMATH

Name: _____

On the _____ day of ChristMATH my math teacher gave to me...
_____

On the _____ day of ChristMATH my math teacher gave to me...
_____

On the _____ day of ChristMATH my math teacher gave to me...
_____

On the _____ day of ChristMATH my math teacher gave to me...
_____

On the _____ day of ChristMATH my math teacher gave to me...
_____

On the _____ day of ChristMATH my math teacher gave to me...
_____

On the _____ day of ChristMATH my math teacher gave to me...
_____

On the _____ day of ChristMATH my math teacher gave to me...
_____

On the _____ day of ChristMATH my math teacher gave to me...
_____

On the _____ day of ChristMATH my math teacher gave to me...
_____

On the _____ day of ChristMATH my math teacher gave to me...
_____

On the _____ day of ChristMATH my math teacher gave to me...
_____

MERRY CHRISTMATH TREES

Please modify directions according to the needs of your students.

PART ONE: Garland Strands (3 total)

Students will use order of operations to make an equation along each garland strand that equals the number in the decagon (star) at the top of their ChristMATH Tree.

Example: 50 goes in the top star

Strand 1: $40 + 10$

Strand 2: $\sqrt{25} \times 4 + 30$

Strand 3: $8^2 \div 2 + \sqrt{225} + 3$

PART TWO: Ornaments (8 total)

Kids put numbers into the ornaments & use order of operations so that it equals the number up in the star. They will have the other students figure out how to make the ornaments equal the star total.

Example: 50 is still in the top star

Numbers in the 8 ornaments:
15   8   20   2   10   6   4   5

Use operations to equal the star up top (solved):
$(15 - 8) \times 20 \div 2 - 10 - (6 - 4) \times 5$

<--

Alternate Example:

Life size ChristMATH tree using bulletin board paper.

MATERIALS NEEDED:

activity sheet

markers, crayons, colored pencils

Completed Example:

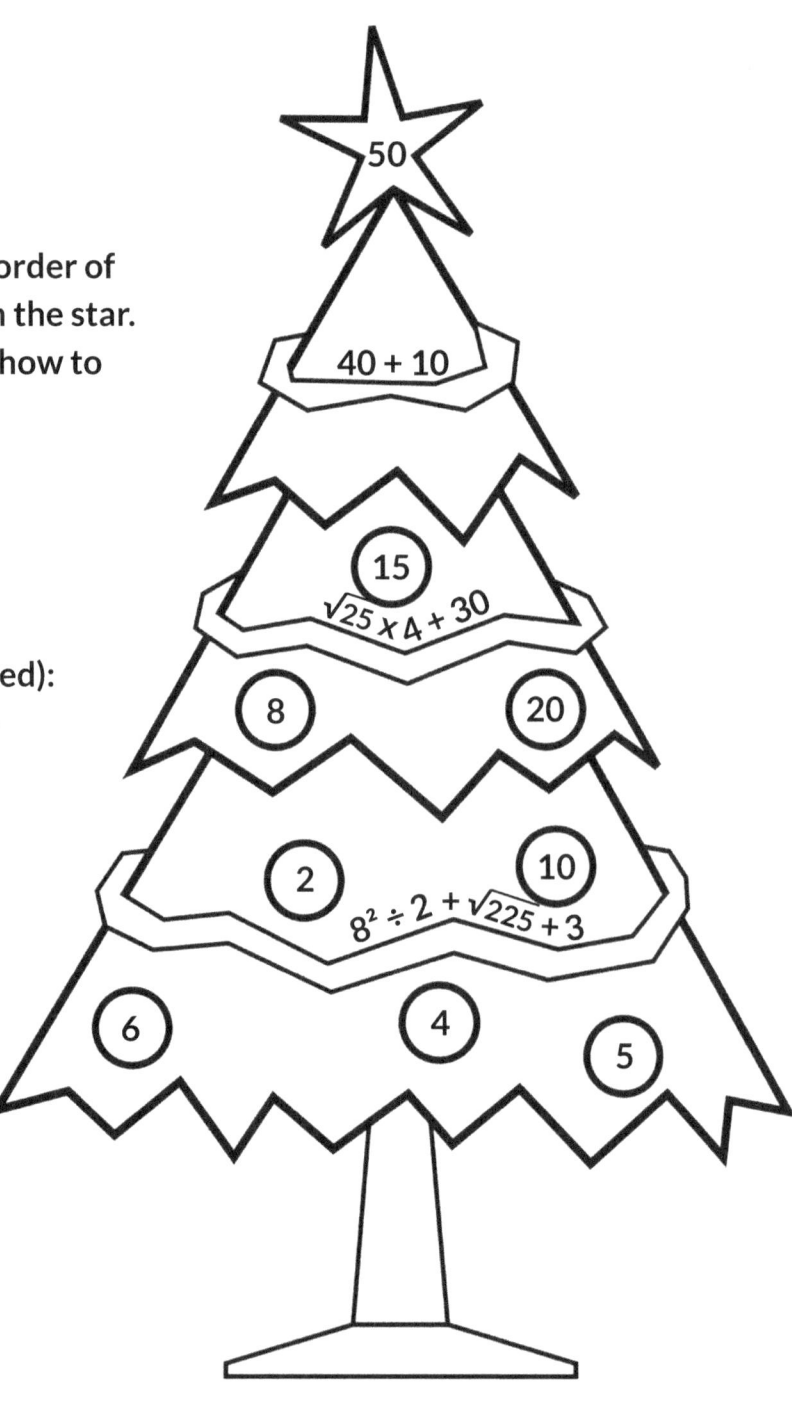

Merry ChristMATH Trees

Name: _____

Directions:

PART ONE: Garland Strands (3 total)

Choose a number and write it in the decagon (star). For each of the 3 garland strands, create a math sentence that equals the number in the decagon (star).

In the garland closest to the decagon, use two different numbers. In the second garland use 3 different numbers without repeating any from the garland above. In the last garland use 4 different numbers. Try to use different numbers for the entire ChristMATH tree!!

Use your awesome math mind to apply square roots and exponents to any numbers in the garland!!

PART TWO: Ornaments (8 total)

Put numbers into the ornaments & use order of operations so that it equals the number up in the star. (Write the solutions on the back side). The other students will try to figure out how to make YOUR ornaments equal the star total!

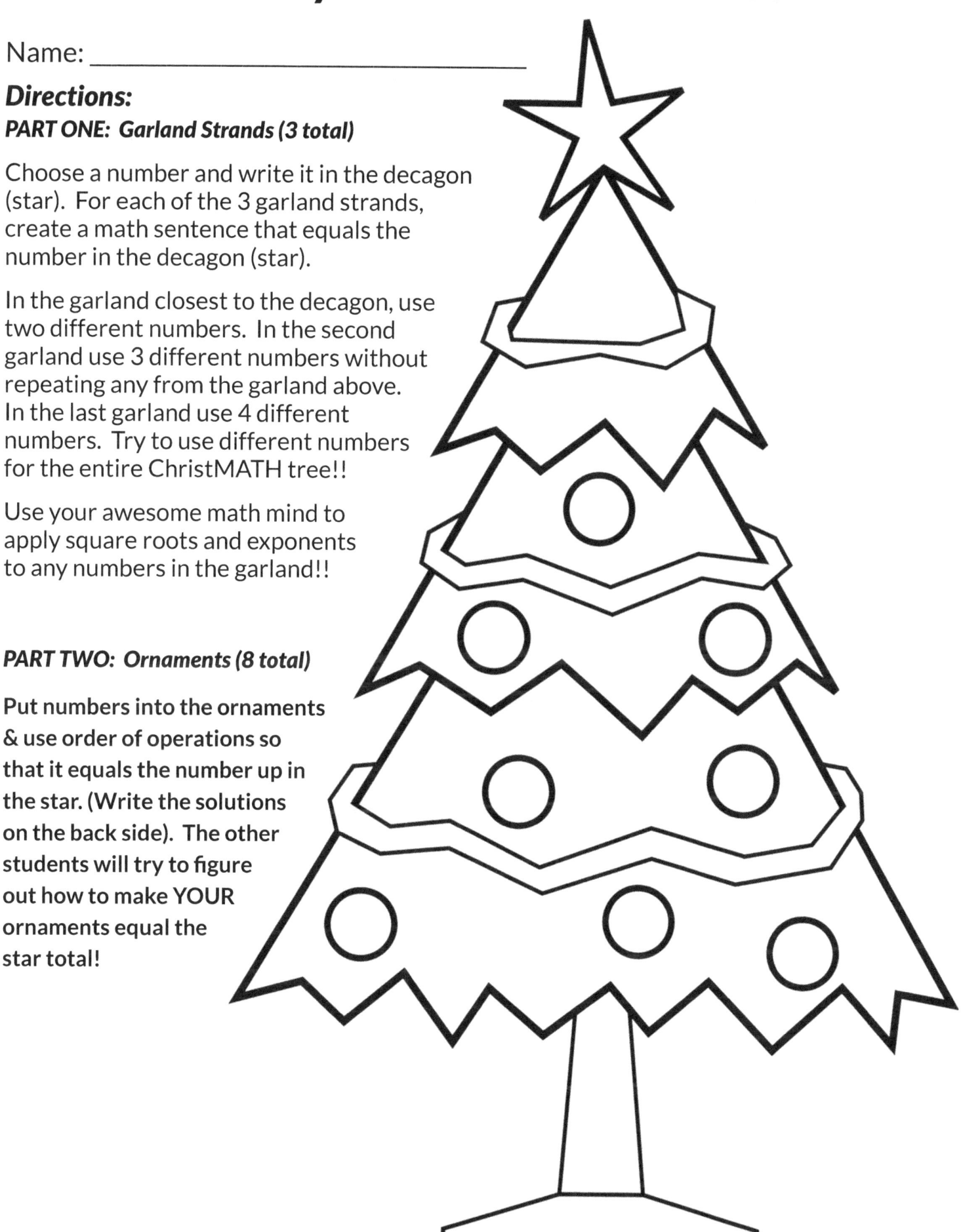

SNOWMATH FLAKEGRAPHS

Students will create snowflakes and circle graphs. Students can create any kind of graph to put into the center of these.

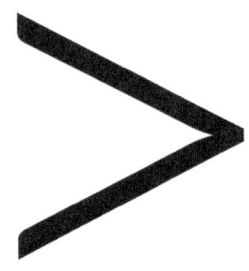

Actual Classroom Example:

MATERIALS NEEDED:

scissors

glue stick

tag board

snowflake directions (find any online)

protractors

pre-made circles or safety compass

a variety of colored paper

SnowMATH Flakegraphs

Name: _____

Directions: You will create three unique snowmath flakegraphs. Please complete this activity sheet BEFORE you create your snowmath flakegraphs. Make sure your percents total 100 BEFORE you fill in the degree and fraction. **You can only use a percent ONCE for this entire project. The percents you use MUST be between 5 and 45.**

1. List 5 things you want for your birthday.   PERCENT   DEGREE   FRACTION

   _____   _____   _____   _____

   _____   _____   _____   _____

   _____   _____   _____   _____

   _____   _____   _____   _____

   _____   _____   _____   _____

Create a snowmath flakegraph and label each sector what you want for your birthday: percent, degree, and fraction. Cut it out and glue it to the center of your snowflake.

2. List 4 characteristics that describe you!!!   PERCENT   DEGREE   FRACTION

   _____   _____   _____   _____

   _____   _____   _____   _____

   _____   _____   _____   _____

   _____   _____   _____   _____

Create a snowmath flakegraph and label each sector your characteristic: percent, degree, and fraction. Cut it out and glue it to the center of your snowflake.

3. List five of your favorite_____   PERCENT   DEGREE   FRACTION

   _____   _____   _____   _____

   _____   _____   _____   _____

   _____   _____   _____   _____

   _____   _____   _____   _____

   _____   _____   _____   _____

Create a snowmath flakegraph and label each sector your favorites: percent, degree and fraction. Cut it out and glue it to the center of your snowflake.

AFRICAN AMERICAN MATHEMATICIANS QUOTE POSTERS

Actual Classroom Example:

MATERIALS NEEDED:

scissors

glue stick

poster board (various colors)

African American Mathematicians QUOTE POSTERS

Name: _____

Directions:

Choose an African American mathematician to research.

Find five or six quotes said by your mathematician.

Print off a picture of your mathematician.

Create different shapes to print & paste (or write) the quotes you choose.

Glue the picture to a piece of poster board. Then glue the different shape quotes around your picture. You will have the opportunity to present these to the class.

Use the worksheet provided to do the following:

List your shapes.

Measure the dimensions using centimeters.

Compute the area and perimeter of each.

Find at least 4 facts about your mathematician that include numbers (dates, ages, etc). Create questions using 4 or more different numbers, and order of operations for others to figure out the numbers.

Please create an answer key and have it ready when you present to the class.

Example: Kimberly Weems was born January 50 – 3 x 10 – 2.

African American Mathematicians QUOTE POSTERS - Research

Your Name: _____

Mathematician's Name: _____

Quotes I Am Using:

1. _____
   _____

2. _____
   _____

3. _____
   _____

4. _____
   _____

5. _____
   _____

6. _____
   _____

African American Mathematicians QUOTE POSTERS - Worksheet

Your Name: _____

Mathematician's Name: _____

Shapes	Dimensions	Area	Perimeter
_____	_____	_____	_____
	_____		
_____	_____	_____	_____
	_____		
_____	_____	_____	_____
	_____		
_____	_____	_____	_____
	_____		
_____	_____	_____	_____
	_____		
_____	_____	_____	_____
	_____		

Four Facts

1. _____
   _____
2. _____
   _____
3. _____
   _____
4. _____
   _____

MATHENTINES 2D VERSION

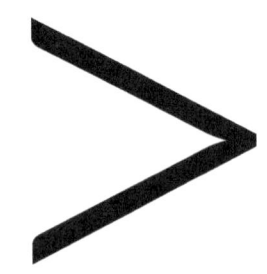

There are three versions of this project. Modify according to the needs of your students. This first version has students problem solve using prior knowledge of two dimensional figures. They create a 2D heart from a circle and a square. The diameter of the circle must equal the side of the square.

See next page for the alternate version (3D MATHentine).

Actual Classroom Examples:

MATERIALS NEEDED:

color paper

safety compass

ruler

glue

markers

TIME:

Each version takes about 80 minutes

MATHentines

2D

Name: _____ (also put your name somewhere on your Mathentine)

Directions:
Using a safety compass to draw a circle, calculate:

	Diameter	Radius	Area	Circumference
Circle	_____	_____	_____	_____

Use a ruler to draw a square that has side lengths which are congruent to the diameter:

	Side Measurements	Area	Perimeter
Square	_____	_____	_____

Figure out how to create a heart with the circle and square. Glue the hearts on paper and label the properties of each shape. Put an "I" next to your heart and something or someone you love after.
Example: I (heart) My Family -or- I (heart) Pizza!

How did you create your heart? _____
_____
_____

How would you compute the area of your heart? _____
_____

Area of Heart _____

How would you compute the distance around heart? _____
_____

Distance Around Heart _____

Does the area change when the two shapes become a heart? _____ because _____
_____

Does the perimeter change when the two shapes become a heart? _____ because _____
_____

MATHENTINES 3D VERSION

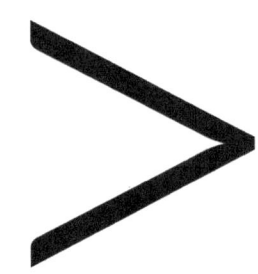

This version has students problem solve using prior knowledge of three dimensional figures and nets. They create a 3-D heart from a nets of a cube and cylinder. Students fill out a project sheet while computing the percent, fraction, and degree of each sector.

See previous page for the first version (2D MATHentine).

Actual Classroom Examples:

MATERIALS NEEDED:

activity sheet with net *(example template supplied on **upcoming** page)*

crayons, markers, colored pencils

tape

TIME:

Each version takes about 80 minutes

MATHentines 3D

Name: _____ (also put your name somewhere on your Mathentine)

A. Decorate all your shapes EXCEPT two adjacent squares. Put an X over the two adjacent squares you are *not* decorating.

B. Using centimeters, measure each dimension to the nearest tenth. When finding area, if necessary, round to the nearest tenth.

1. Side of a Square _____   Area of Square _____   Cube Surface Area _____

How did you compute the surface area of the cube? _____

2. Rectangle Length _____   Width _____   Area of Rectangle _____

How many congruent rectangles are there? _____   Total Rectangle Area _____

3. Diameter _____   Radius _____   Area of Circle _____

How many congruent circles are there? _____   Total Circle Area _____

4. What is the surface area of your heart? _____

5. How did you compute the surface area of your heart? _____
_____

6. Just for fun.... What is the volume of the cube? _____

7. What figure could the rectangle and circles create? _____   Volume _____

8. Determine total volume of your heart: _____

Directions to create your heart:
Cut out the T (net for the cube). Create a cube and tape it together.
Cut out the two rectangles and circles. Cut the circles in half.
Figure out how to tape the pieces together to create a HEART!

May the L♥ve of Math fill your heart!! Happy Mathentine's Day!

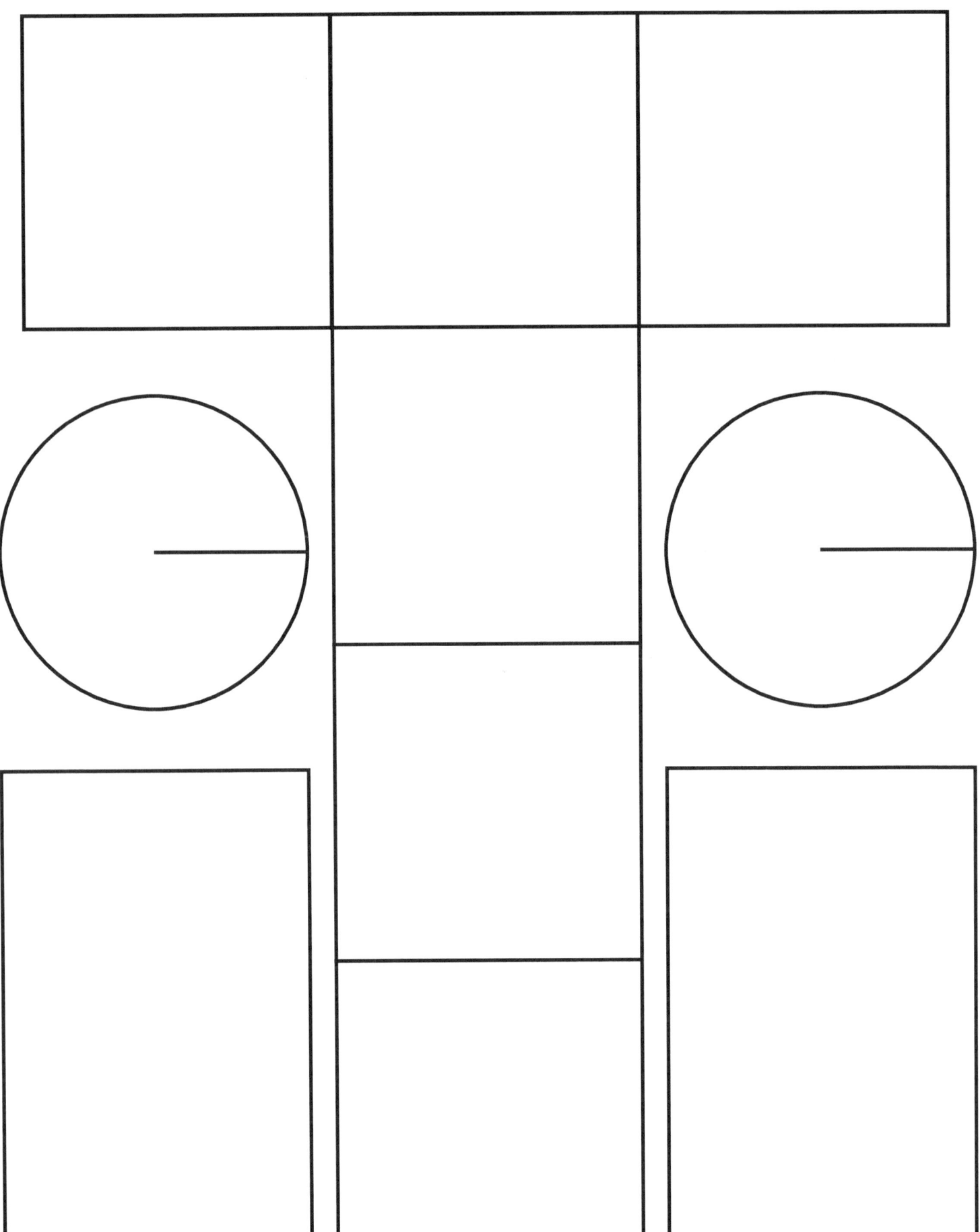

MATHentines

Alternate Version

Name: _____ (also put your name somewhere on your Mathentine)

Directions:
Create any polygon. List the shapes created. Measure the dimensions of each shape. Put the shapes together as a collage heart. You may create more than one heart.

Write "I" above it and something or someone you love below it. *Example: I (heart) Math!*

Polygon	Dimensions
_____	_____
_____	_____
_____	_____
_____	_____
_____	_____
_____	_____
_____	_____
_____	_____
_____	_____
_____	_____
_____	_____
_____	_____
_____	_____

I 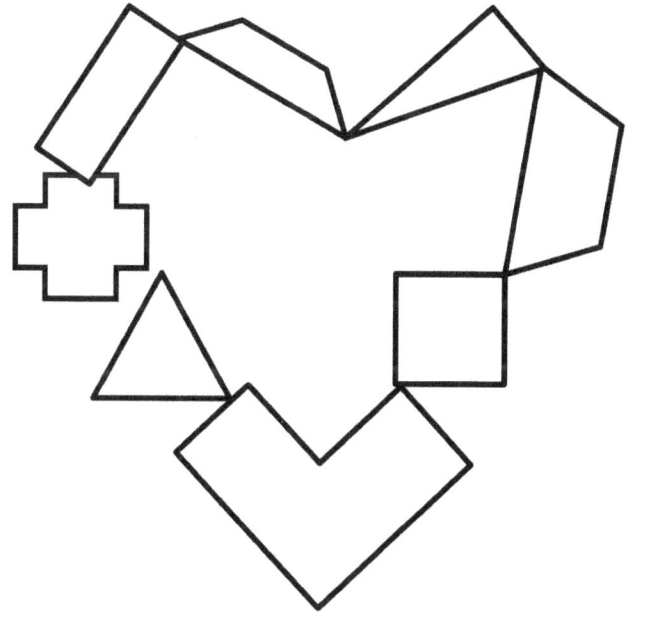 Math

mathEGGtastic hunt

Students will create math problems according to the standards/concepts for others to answer. Modify according to grade level.

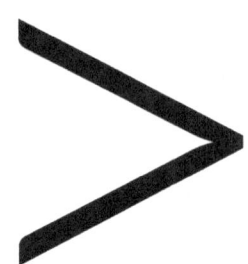

MATERIALS NEEDED:

activity sheet to create math problems

bucket for each group to collect eggs

5 or more different colored plastic eggs

TIME / PLANNING:

Kids love this!! It takes 40 minutes for them to create questions. I got help from other teachers on their planning time to put eggs outside. I have several classes so I labeled the eggs the hour I had each class. The class was informed only to choose eggs that were their designated color and had their hour on it. It is fun to watch them search!

MathEGGtastic Hunt

Name: _____

Directions: Have each group of students create math problems in the sections on the paper given. Cut out each problem and put a problem in each different colored plastic egg. Each group is a different color and can only find *their* color eggs. The first group to find all their eggs and answer the most problems correctly wins!

Area	Order of Operation
Perimeter	Compute Missing Angle
Combine Like Terms	Solve An Equation

MATHLICIOUS GAMES

$$\frac{\text{Fun} + \text{Math}}{\text{LEARN}}$$

I try to allow my students to play part of one of these each day. They are learning while having fun. My students turn these in for credit. Everything in my class has value. These games can be differentiated to meet the needs of all students. These can be used for classroom management.

Helpful Hint: When students are using dice have them roll on a paper plate.
This helps keep the dice on the desk.

YAHMATHZEE

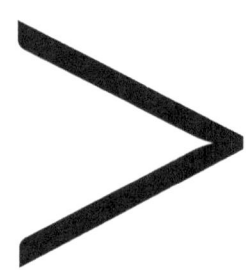

This can be a whole class activity and groups play against each other, or individually.

This can be used any time and many times throughout the year. There is a blank copy for you to create and adapt to fit the needs of your students. You don't have to play an entire game all in one day. You can spread it out for several days.

Here are the rules you can put on the games you create:

Rules –

Roll the dice twice to get the dimensions of a rectangle OR base and height of a triangle. Draw the rectangle or triangle and label its dimensions. Compute the area. If category requires perimeter of a triangle, students will choose a number that follows the rule for creating a triangle. If your area and/or perimeter match the category, enter the area as your score. If, on your turn, what you roll does not fit the category, put a 0 for your score. On the FREEBIE turn you may choose one category that received a 0 and try again. Everyone gets 10 turns. Total your score. Highest score wins!!!

There are many ways to tweak the directions to fit the needs of students. For example, I have all the students begin with category 1, then 2, then 3 etc… After they are comfortable with playing, they can roll the dice and then pick a category that fits what they rolled.

3 Scoresheets are provided on the following pages:

- Basic Math Concepts
- Algebra
- Blank (Let Students Create Their Own)

MATERIALS NEEDED:

dice or spinners

If playing as a whole class, I use dice online.

YahMATHzee!

Put the initials of each player in the columns – *circle your own initials.*

Category	Player ___	Player ___	Player ___	Player ___	Player ___	Player ___	Player ___	Player ___
1) Area is a multiple of 5								
2) Perimeter is a factor of 100								
3) A + P = even number								
4) P – A = negative integer								
5) Perimeter + 7 is a prime number								
6) Area < Perimeter								
7) P + A is divisible by 4								
8) Area OR Perimeter is a perfect square								
9) A – P + 9 = odd number								
10) FREEBIE								
Total Points								

THIS ACTIVITY CAN BE PLAYED WITH ANY SIDED DICE OR SPINNERS!!!

Game Rules: Roll the dice twice to get the dimensions of a rectangle OR base and height of a triangle. Draw the rectangle or triangle and label its dimensions. Compute the area. If category requires perimeter of a triangle, students will choose a number that follows the rule for creating a triangle. If your area and/or perimeter match the category, enter the area as your score. If, on your turn, what you roll does not fit the category, put a 0 for your score. On the FREEBIE turn you may choose one category that received a 0 and try again. Everyone gets 10 turns. Total your score. Highest score wins!!!
The teacher may adjust the game & explain alternate rules for this worksheet.

YahMATHzee!

Put the initials of each player in the columns – *circle your own initials.*

Category	Player ___	Player ___	Player ___	Player ___	Player ___	Player ___	Player ___	Player ___
1) area is a multiple of the slope $y = 3x + 8$								
2) y-intercept is a factor of the perimeter $y = x + 4$								
3) slope + area = even number $y = 5x + 7$								
4) y-intercept x area = odd number $y = 7x + 11$								
5) the x coordinate is a factor of the area (2, 5)								
6) y coordinate + perimeter = prime number (6, 12)								
7) the difference between x value and area = composite number (15, 7)								
8) the sum of the x and y coordinate is a multiple of the area (3, 2)								
9) area - perimeter is a factor of the y-intercept $y = 8x + 24$								
10) FREEBIE								
Total Points								

THIS ACTIVITY CAN BE PLAYED WITH ANY SIDED DICE OR SPINNERS!!!

Game Rules: Roll the dice twice to get the dimensions of a rectangle OR base and height of a triangle. Draw the rectangle or triangle and label its dimensions. Compute the area. If category requires perimeter of a triangle, students will choose a number that follows the rule for creating a triangle. If your area and/or perimeter match the category, enter the area as your score. If, on your turn, what you roll does not fit the category, put a 0 for your score. On the FREEBIE turn you may choose one category that received a 0 and try again. Everyone gets 10 turns. Total your score. Highest score wins!!!
The teacher may adjust the game & explain alternate rules for this worksheet.

YahMATHzee!

Put the initials of each player in the columns – *circle your own initials.*

Category	Player ___	Player ___	Player ___	Player ___	Player ___	Player ___	Player ___	Player ___
1)								
2)								
3)								
4)								
5)								
6)								
7)								
8)								
9)								
10) FREEBIE								
Total Points								

THIS ACTIVITY CAN BE PLAYED WITH ANY SIDED DICE OR SPINNERS!!!

Game Rules: Roll the dice twice to get the dimensions of a rectangle OR base and height of a triangle. Draw the rectangle or triangle and label its dimensions. Compute the area. If category requires perimeter of a triangle, students will choose a number that follows the rule for creating a triangle. If your area and/or perimeter match the category, enter the area as your score. If, on your turn, what you roll does not fit the category, put a 0 for your score. On the FREEBIE turn you may choose one category that received a 0 and try again. Everyone gets 10 turns. Total your score. Highest score wins!!!
The teacher may adjust the game & explain alternate rules for this worksheet.

MATHOGGLE

This mathtastic game will boggle your students' minds! They must create math sentences, using numbers adjacent or diagonal to each other which equal the target number you give them. I prefer they use more than one operation to arrive at the target number.

Modify according to grade level.

The spaces are blank. I recommend you make one copy, fill in your own numbers & then duplicate that for your class.

EXAMPLE:

Target Number = 50

13 X 3 + 2 + 9 = 50

6	13	20	5
18	3	46	25
34	2	10	4
8	40	9	16

Mathoggle!

Name: _____

Directions: Please list below the math sentences you create. Numbers used must be adjacent or vertical from the previous number. You can only use a number ONCE in each sentence. You must have 3 or more numbers to create a Mathoggle. Use two or more different operations.

1) _____

2) _____

3) _____

4) _____

5) _____

6) _____

7) _____

8) _____

9) _____

10) _____

EQUATIONANZA

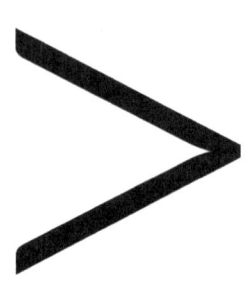

This game can be used many different ways and adapted to any grade level. This is also collected for credit. You can choose kids to shout out numbers to use, or pick numbers from a bucket or make cards with numbers on them. I usually use five numbers and have students work in pairs or groups. Use what works for you!! Whatever group or pair creates the most equations wins! Students love to share these.

Students record the five numbers being used. They must use each number once to create an equation that equals the EquationANZA number. The EquationANZA number is the date, or may be chosen by teacher or student. I usually use the date. I set a timer for as long as I want to give them, usually 5 – 10 minutes. After the timer goes off, we share our equations. You could also have the students shout "EquationANZA" when they create their first equation.

Actual Classroom Example:

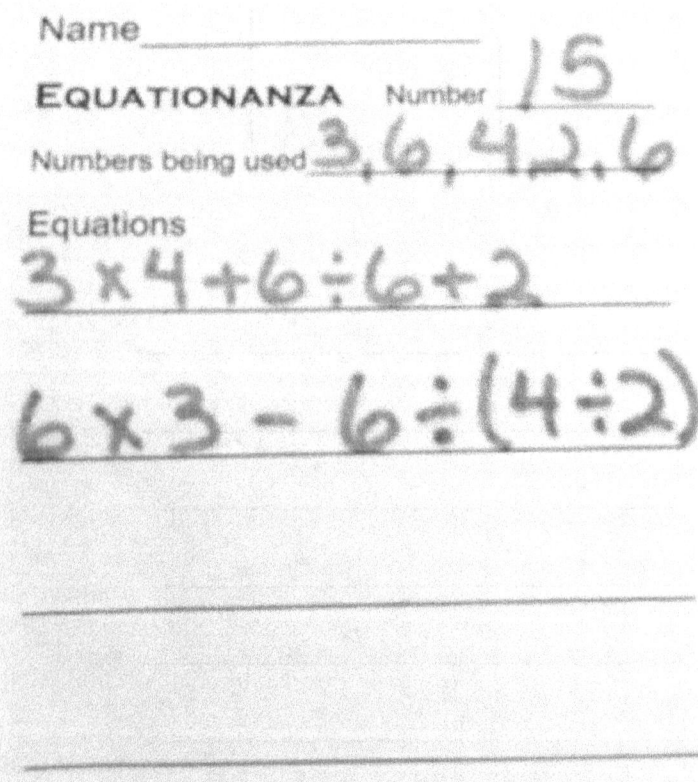

EquationANZA

Name _____
EquationANZA Number _____
Numbers Being Used _____
Equations:

_____
_____
_____
_____

Name _____
EquationANZA Number _____
Numbers Being Used _____
Equations:

_____
_____
_____
_____

✂ - - - - - - - - - - -

Name _____
EquationANZA Number _____
Numbers Being Used _____
Equations:

_____
_____
_____
_____

Name _____
EquationANZA Number _____
Numbers Being Used _____
Equations:

_____
_____
_____
_____

Name _____
EquationANZA Number _____
Numbers Being Used _____
Equations:

_____
_____
_____
_____

Name _____
EquationANZA Number _____
Numbers Being Used _____
Equations:

_____
_____
_____
_____

DYNAMATHIC DICE DILEMMA

I play this as a class. The game spaces create "one number". For instance, if they fill in 5, 2, 6, 1, and 3 for Game 1, their final number is 52,613. ***Note you can add commas or decimals to adapt to the needs of your students.*** Class decides <u>before the game starts</u>, whether the greatest or lowest value is the winner. You can play as many games as time allows. This is a mathabulous game when you have 5 minutes or so before class is done. Kids keep this in their math binders or folders. You can use 6 – 10 sided dice. Most ten sided dice have a 0 instead of 10.

I have also played where the students have to have a number higher than mine or exactly mine in order to win. It's fun either way! Kids will ask to play, and I respond, "If we get our lesson done." When it is complete, I collect it.

Each game has one triangle spot. If students' number in their triangle matches what the teacher put, they win the "triangle treat" (teacher's discretion).

Alternate Version

Another version of DynaMATHic Dice… I play this after each journal entry.

Have students draw six squares like this:

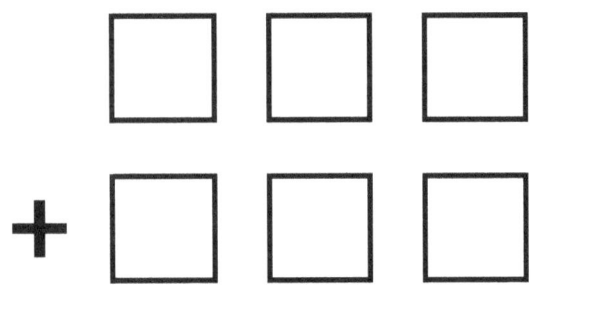

Use interactive dice or dice in a paper plate. Roll six times. After each roll, students should put a number in one of their squares. After six squares are filled in, students should find the sum. Have numbers (100 to 1200 in multiples of 100) in a bucket for someone to pick out and then the group or student who is closest to that number wins.

DynaMATHic Dice Dilemma

Name: _____

Directions: Decide as a class if the greatest or lowest value is the winner. Then ROLL AWAY!! Everyone must record the number rolled before the next rolling of the dice.

Game 1: ☐ ☐ △ ☐ ☐

Game 2: ☐ ☐ ☐ △ ☐

Game 3: ☐ ☐ ☐ ☐ △ ☐

Game 4: ☐ △ ☐ ☐ ☐ ☐

Game 5: △ ☐ ☐ ☐ ☐ ☐ ☐

Game 6: ☐ ☐ ☐ ☐ ☐ △ ☐ ☐

Game 7: ☐ ☐ ☐ △ ☐ ☐ ☐ ☐

Game 8: ☐ ☐ ☐ ☐ △ ☐ ☐ ☐

Game 9: ☐ ☐ △ ☐ ☐ ☐ ☐ ☐ ☐

Game 10: ☐ ☐ ☐ ☐ ☐ △ ☐ ☐ ☐

Game 11: ☐ ☐ ☐ △ ☐ ☐ ☐ ☐ ☐

BaMATHanana

Have you seen the letters that come in bags shaped like a banana? I give these out to pairs of students and they create a crossword using math terms. This may be adapted to younger grades where the letters don't have to connect with each other. They copy their BaMATHanana on the grid paper. I also have students share these with the class. Then they choose three words, and on the back of the grid paper, they prove they know the meaning of those words. I have several words in a bucket. I pick a word out and any students who have that word as a part of their BaMATHanana win. The pair with the most words can also win.

Actual Classroom Examples:

BaMATHanana

Name: _____

Directions: Copy your BaMATHanana game words onto the graph paper. On the backside, choose $\sqrt{9}$ words and prove you know the meaning. Draw a picture, write an example, or use it in a sentence.

BinGRAPHo

This game is a mathabulous way to practice ordered pairs. I created two versions of this game. I created it to help students understand the coordinate plane and where to plot points.

Version One:

Students plot ten points using a marker and then when I call out ordered pairs they use another color marker to plot the points I'm calling out. They put an X over any of the ten points they plotted if I call it out. When they get three X's anywhere they call out BinGRAPHo! Remember this can be adapted to fit the needs of your students. You differentiate by having students plot points only in certain quadrants until they are comfortable with all four quadrants.

Version Two:

The second version already has linear equations that I have graphed FOR YOU on the coordinate plane. Each BinGRAPHo card is different, and you distribute them, similar to normal "bingo cards".

Following the BinGRAPHo card templates provided, you will find the ordered pairs that match the points plotted on the graph.

As I call out any of the ordered pairs on the list, students must plot the points. If I call out an order pair that is on a line, they put an X. When a student has four X's on one line, they call out BinGRAPHo!

In my classroom, these activities are collected for credit.

BinGRAPHo

Version One

Name: _____

Directions: Using a marker, plot 10 points on the graph. Do not plot a point greater than 4. When your coordinate is called out, use a different color marker to put an X over your plotted point. Call out BinGRAPHo when you have three X's anywhere on your game board!

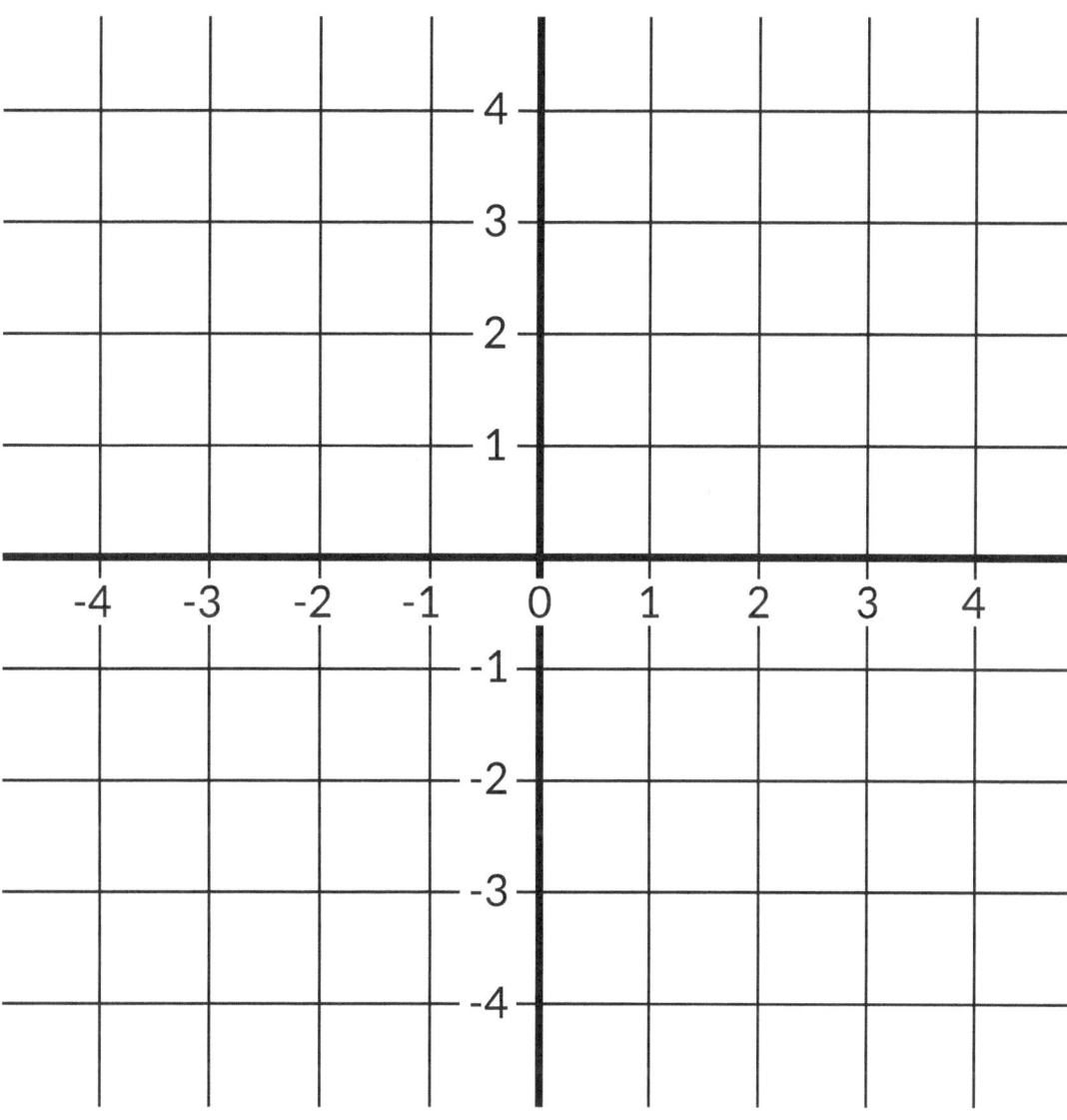

BINGRAPHO

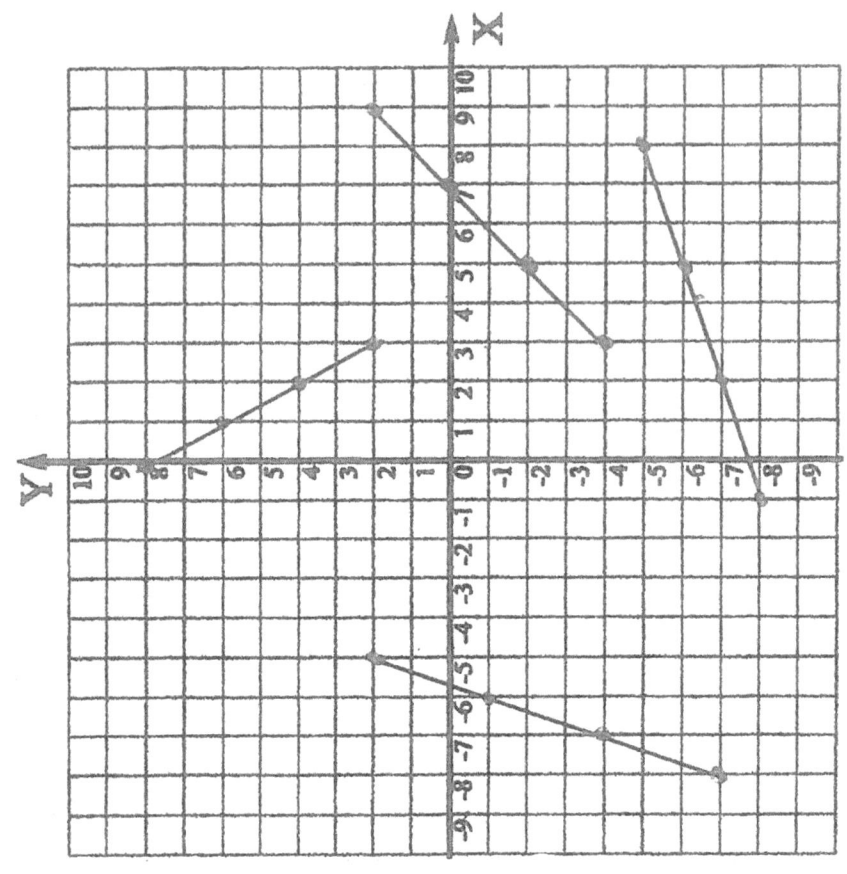

Plot the ordered pairs being called out by putting an X over the point. When you have all four X's along any one line, call out BinGRAPHo!

BINGRAPHO

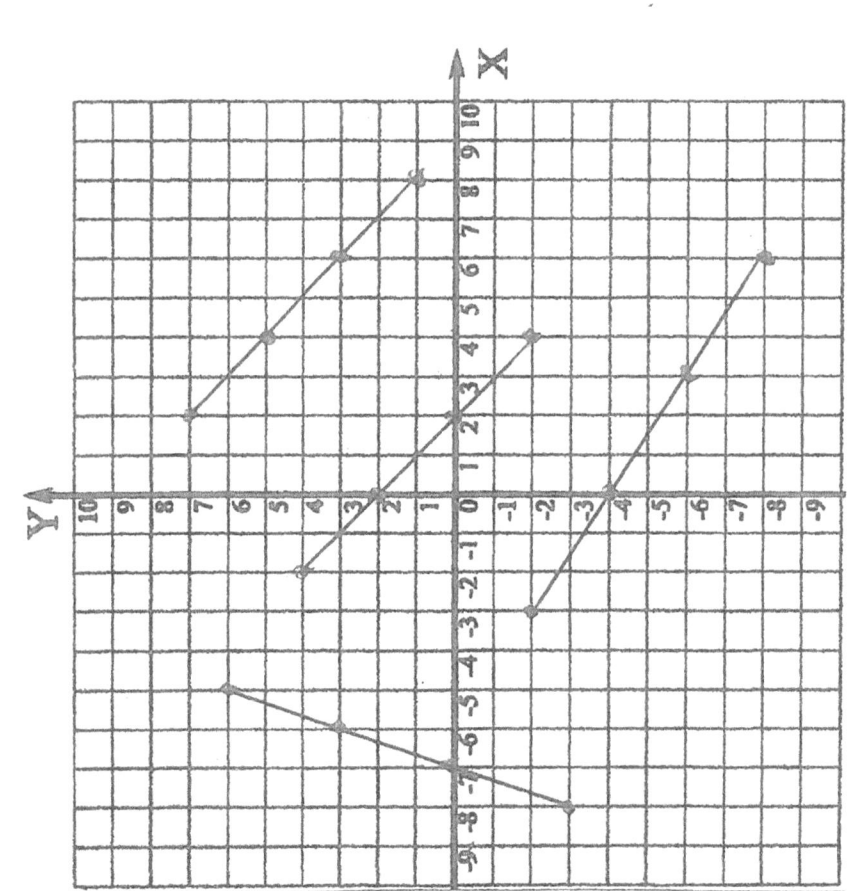

Plot the ordered pairs being called out by putting an X over the point. When you have all four X's along any one line, call out BinGRAPHo!

BINGRAPHO

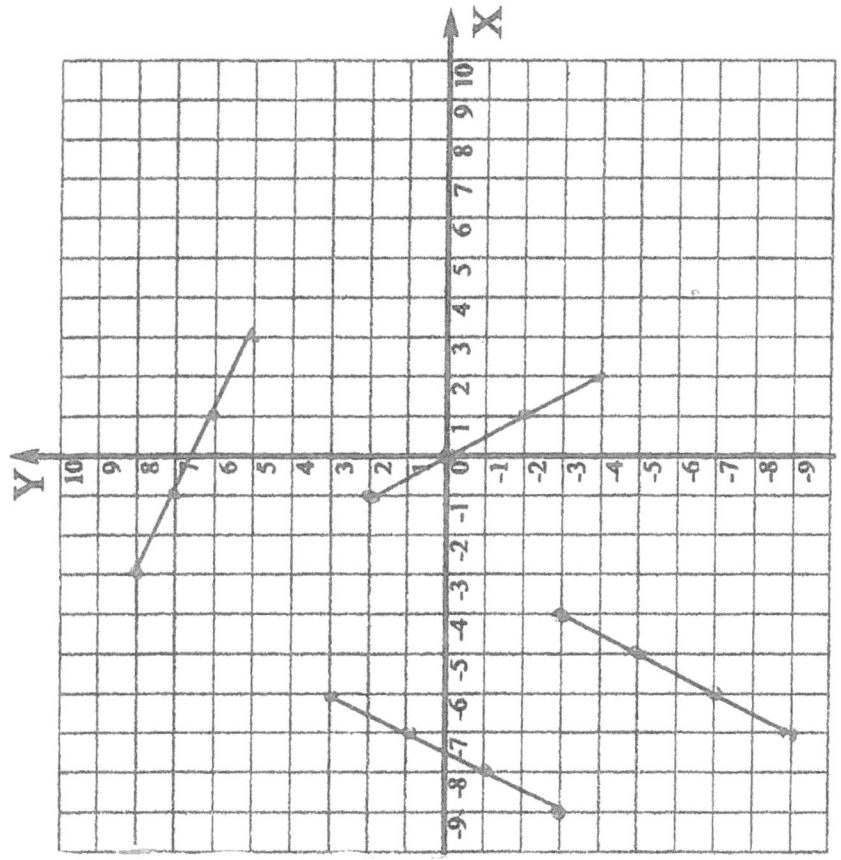

Plot the ordered pairs being called out by putting an X over the point. When you have all four X's along any one line, call out BinGRAPHo!

BINGRAPHO

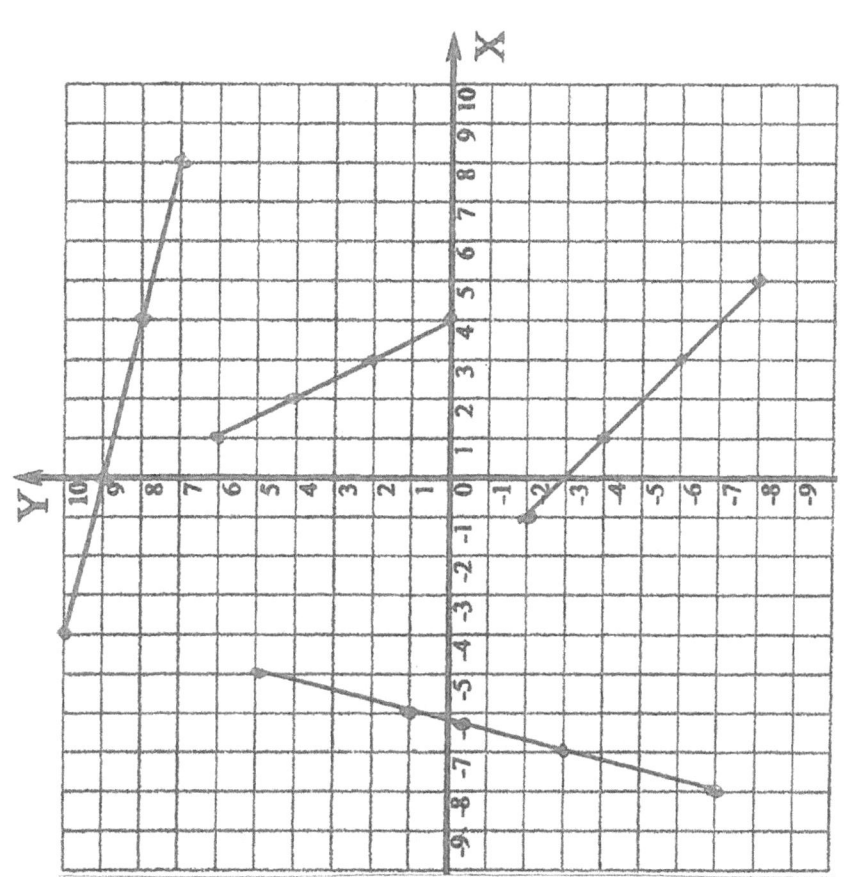

Plot the ordered pairs being called out by putting an X over the point. When you have all four X's along any one line, call out BinGRAPHo!

BINGRAPHO

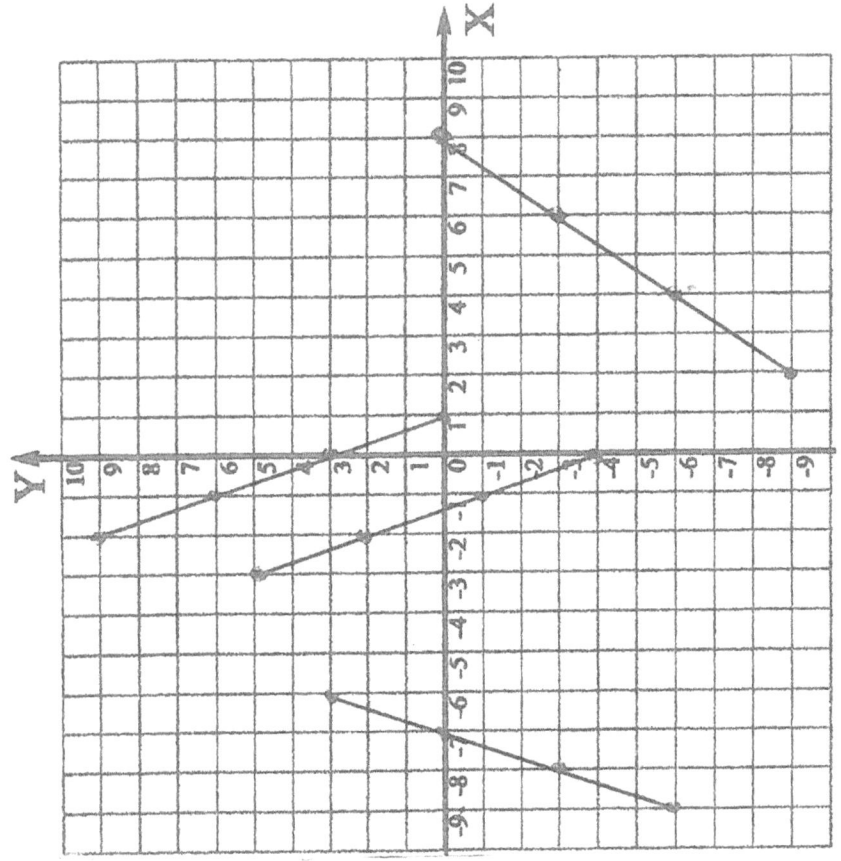

Plot the ordered pairs being called out by putting an X over the point. When you have all four X's along any one line, call out BinGRAPHo!

BINGRAPHO

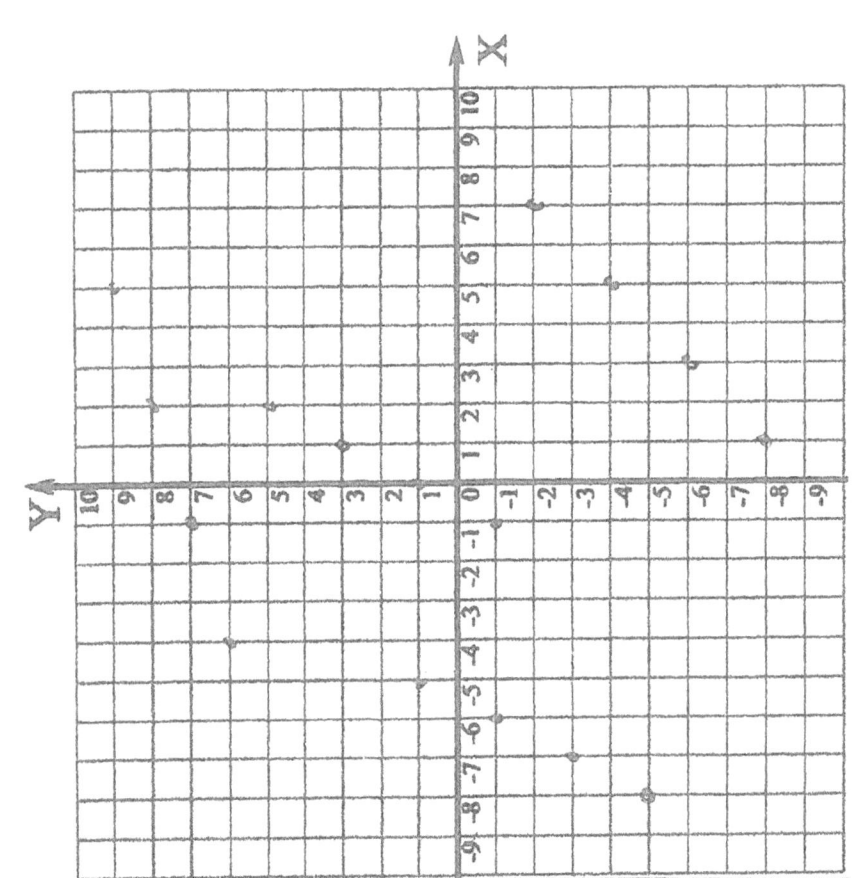

Plot the ordered pairs being called out by putting an X over the point. When you have all four X's along any one line, call out BinGRAPHo!

BINGRAPHO

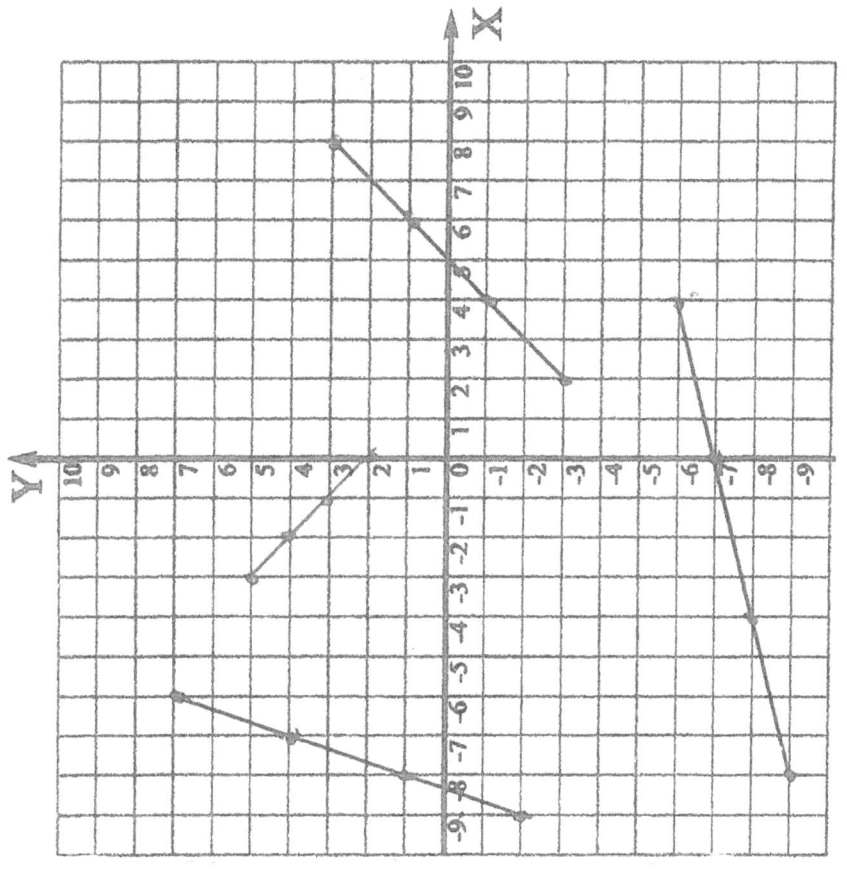

Plot the ordered pairs being called out by putting an X over the point. When you have all four X's along any one line, call out BinGRAPHo!

BINGRAPHO

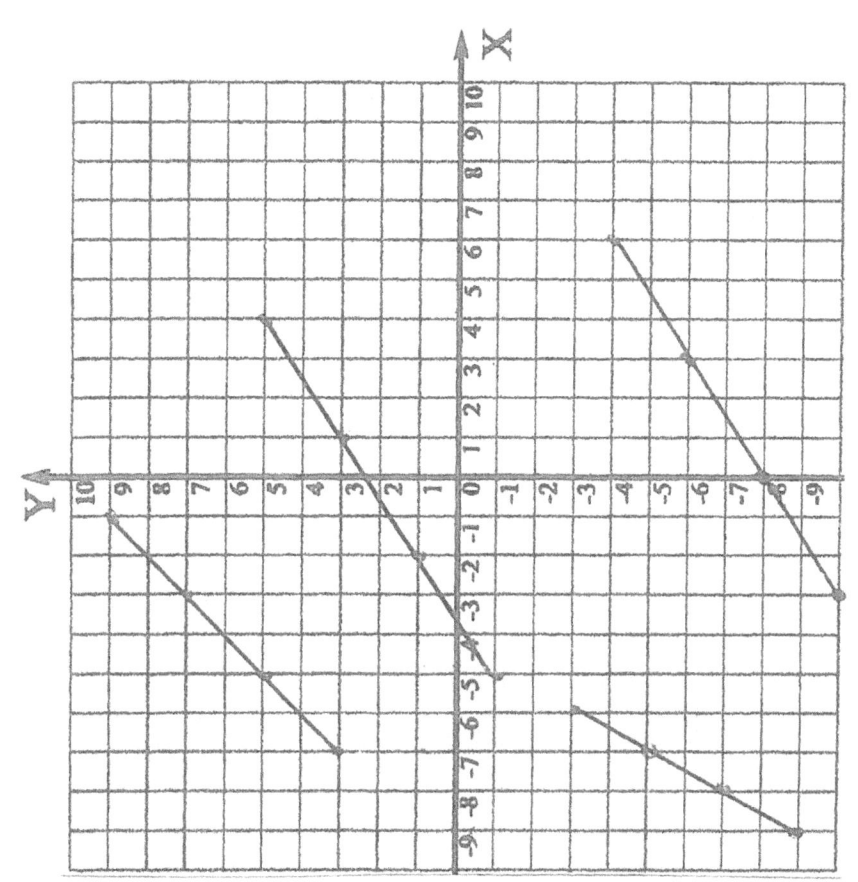

Plot the ordered pairs being called out by putting an X over the point. When you have all four X's along any one line, call out BinGRAPHo!

BINGRAPHO

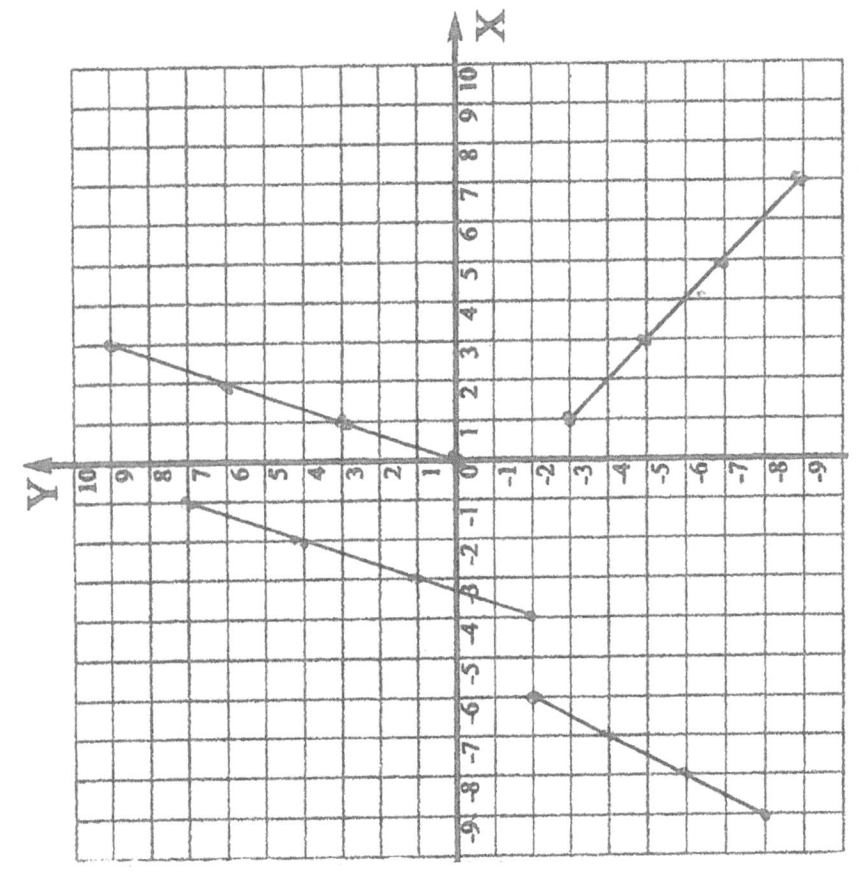

Plot the ordered pairs being called out by putting an X over the point. When you have all four X's along any one line, call out BinGRAPHo!

BINGRAPHO

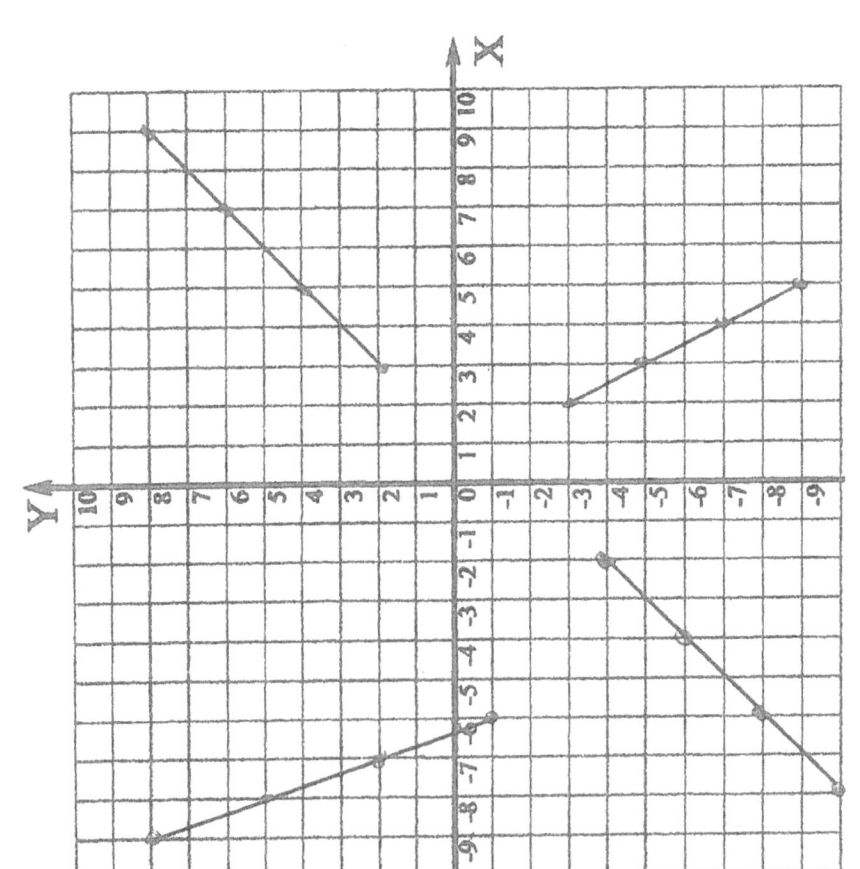

Plot the ordered pairs being called out by putting an X over the point. When you have all four X's along any one line, call out BinGRAPHo!

BINGRAPHO

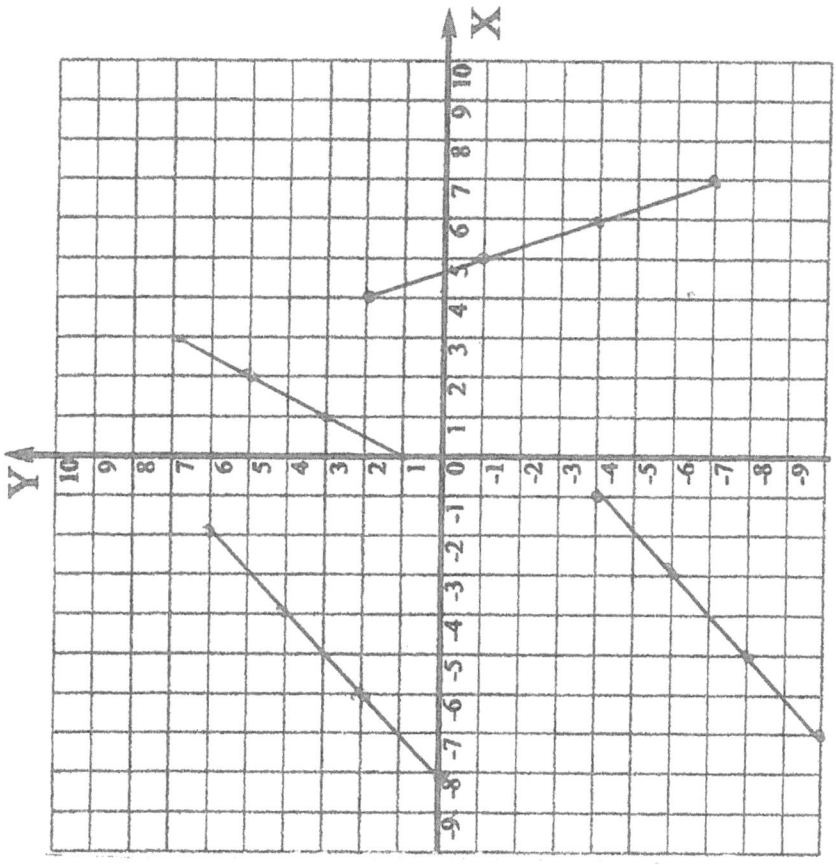

Plot the ordered pairs being called out by putting an X over the point. When you have all four X's along any one line, call out BinGRAPHo!

BINGRAPHO

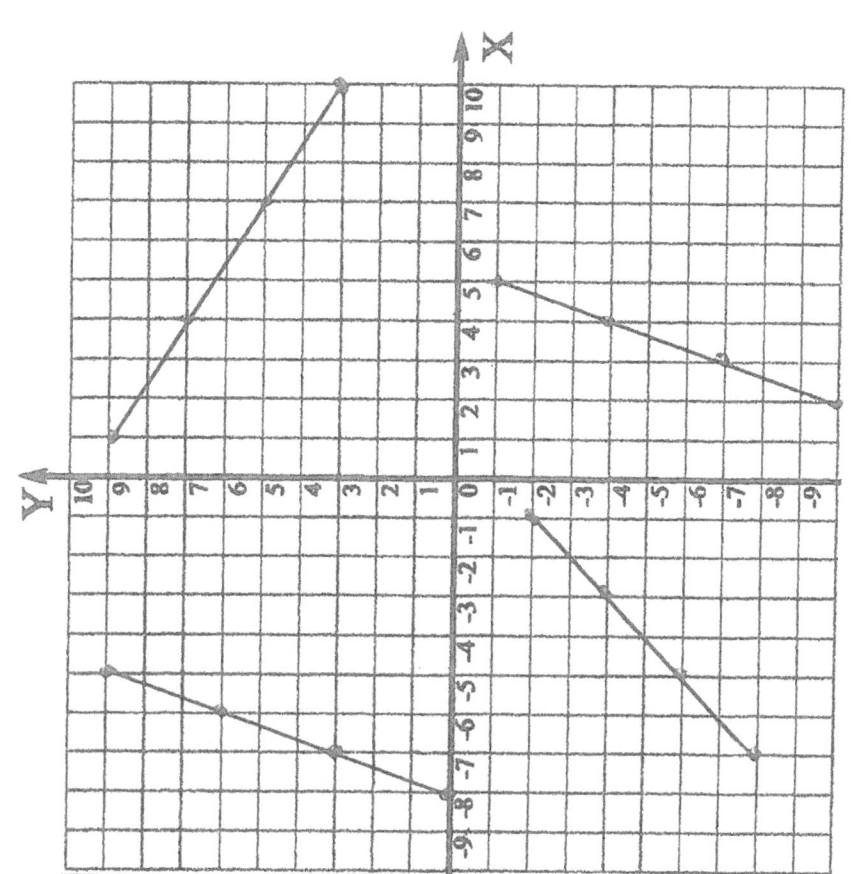

Plot the ordered pairs being called out by putting an X over the point. When you have all four X's along any one line, call out BinGRAPHo!

BINGRAPHO

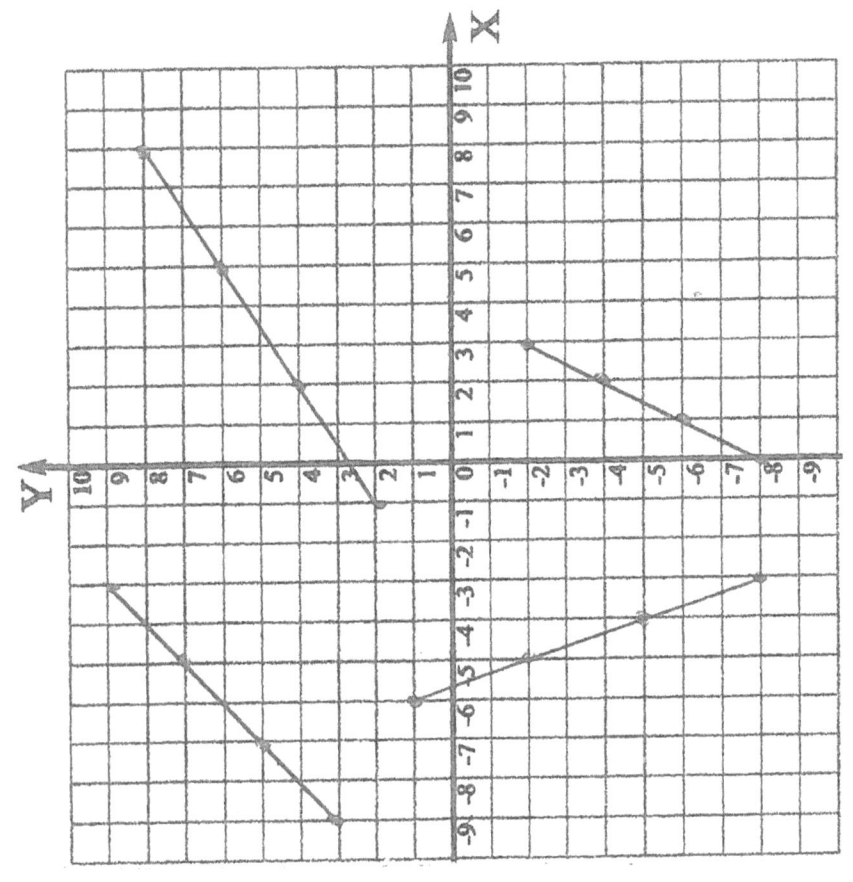

Plot the ordered pairs being called out by putting an X over the point. When you have all four X's along any one line, call out BinGRAPHo!

BINGRAPHO

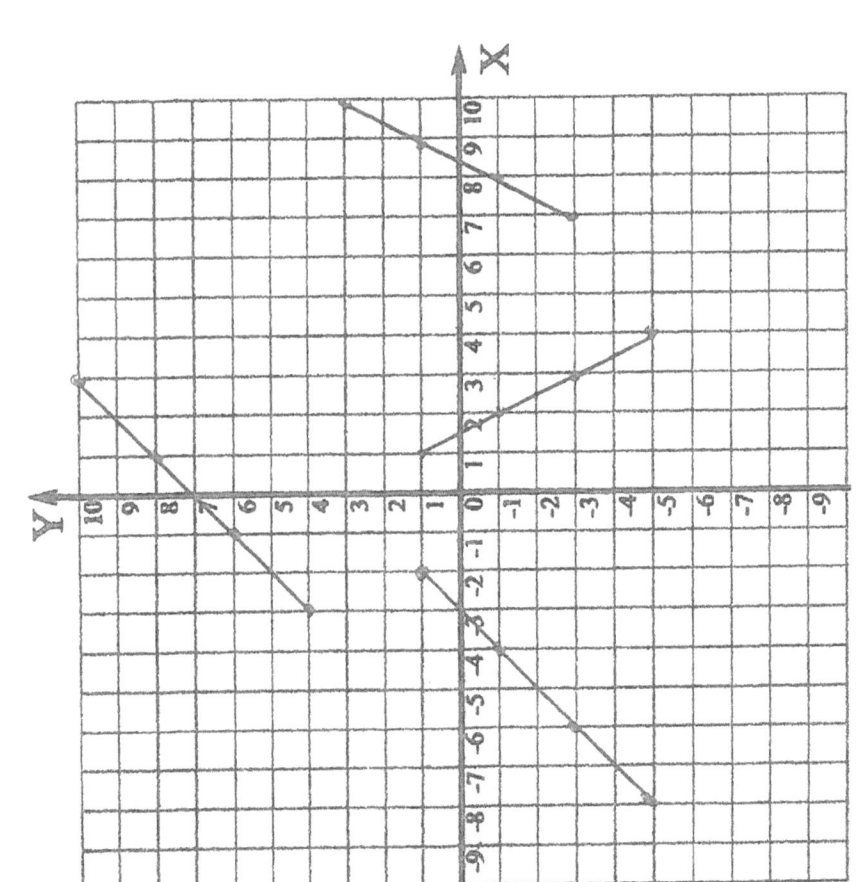

Plot the ordered pairs being called out by putting an X over the point. When you have all four X's along any one line, call out BinGRAPHo!

BINGRAPHO

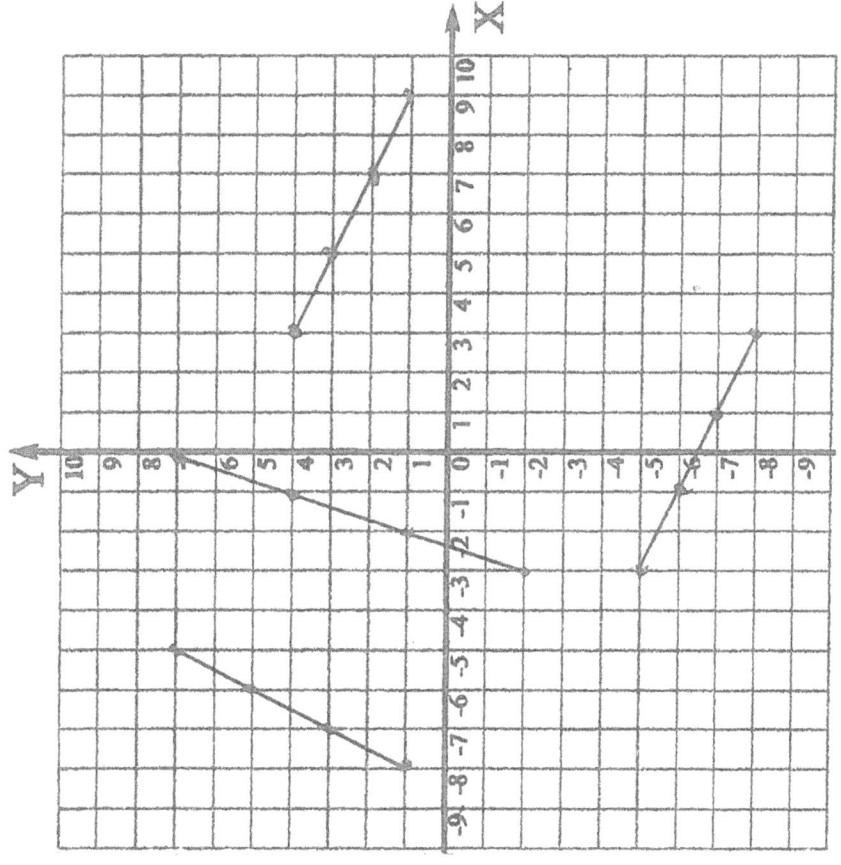

Plot the ordered pairs being called out by putting an X over the point. When you have all four X's along any one line, call out BinGRAPHo!

BINGRAPHO

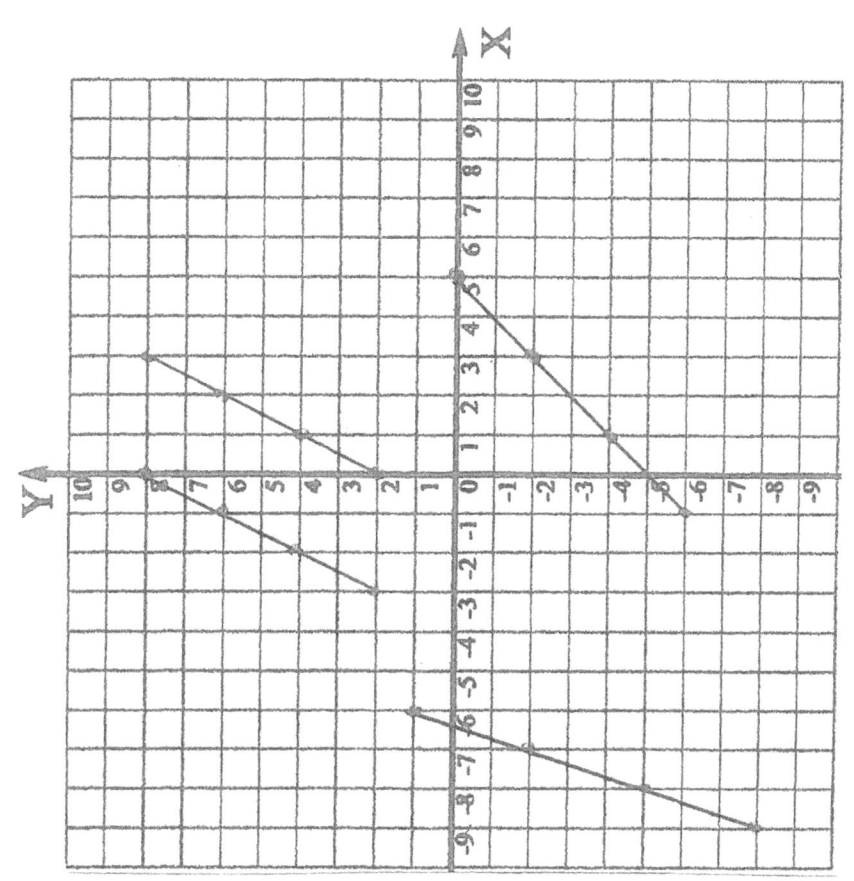

Plot the ordered pairs being called out by putting an X over the point. When you have all four X's along any one line, call out BinGRAPHo!

BINGRAPHO

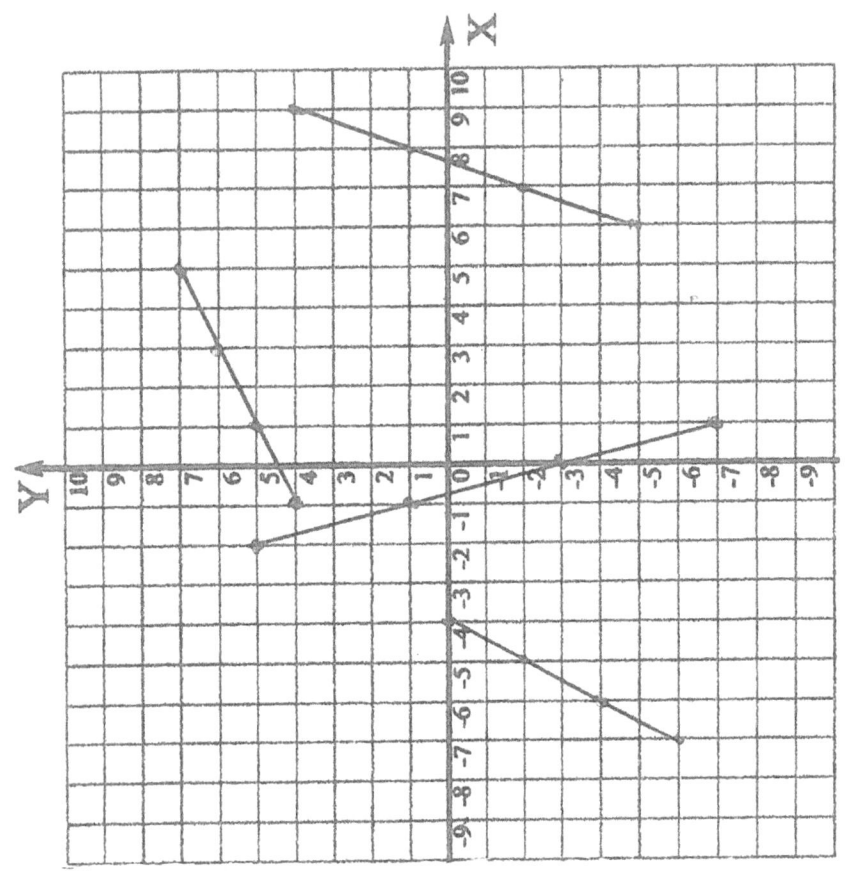

Plot the ordered pairs being called out by putting an X over the point. When you have all four X's along any one line, call out BinGRAPHo!

BINGRAPHO

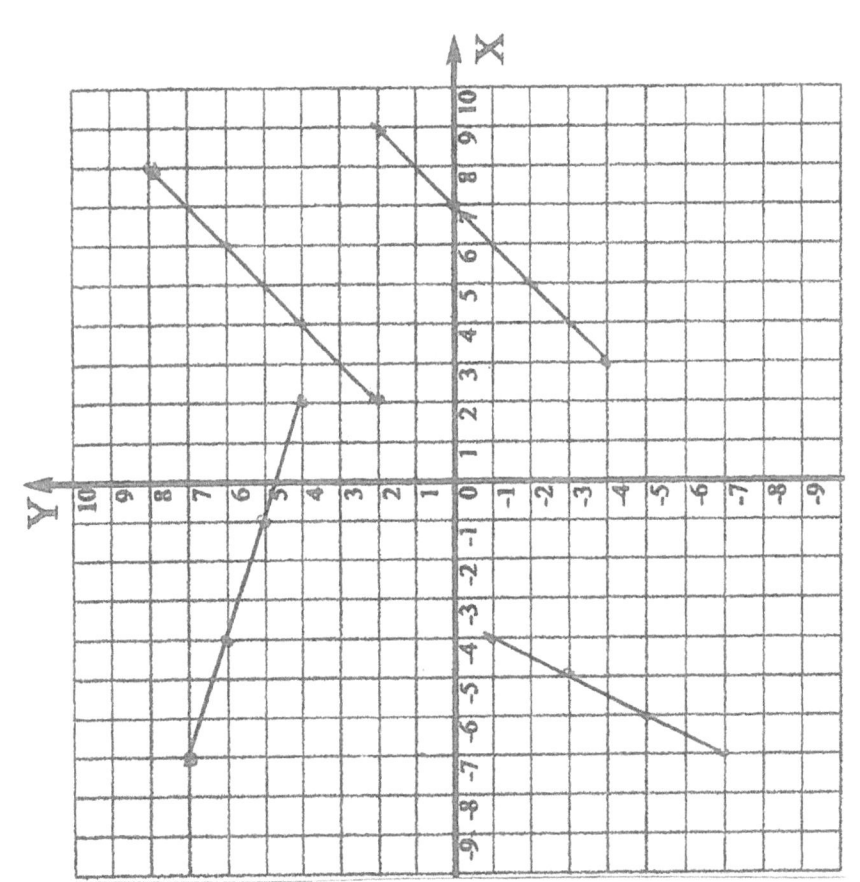

Plot the ordered pairs being called out by putting an X over the point. When you have all four X's along any one line, call out BinGRAPHo!

BINGRAPHO

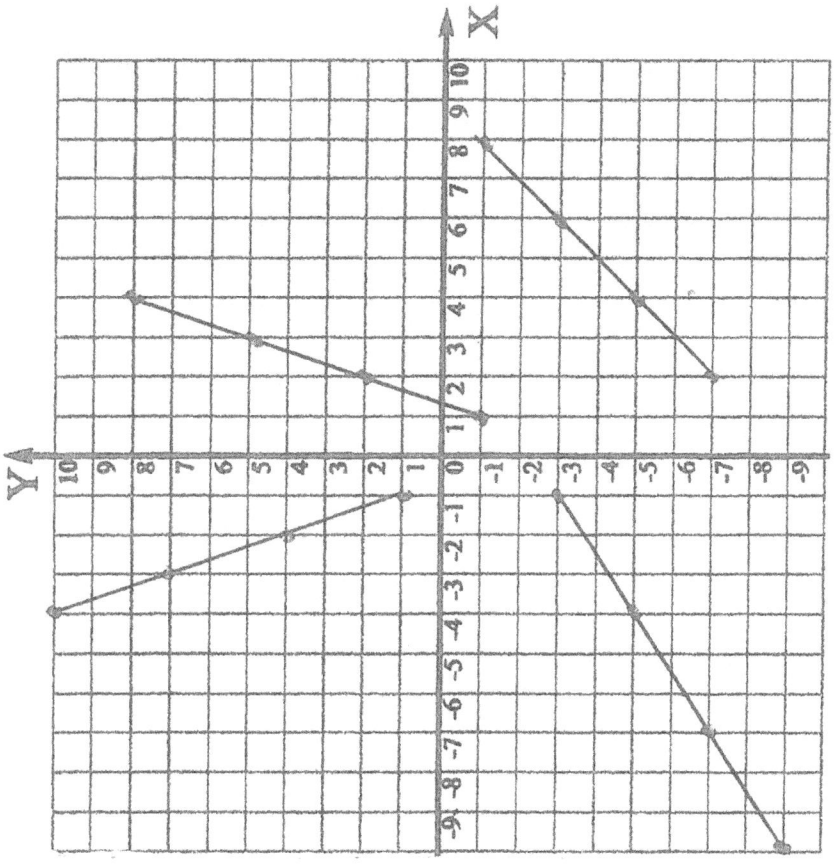

Plot the ordered pairs being called out by putting an X over the point. When you have all four X's along any one line, call out BinGRAPHo!

BINGRAPHO

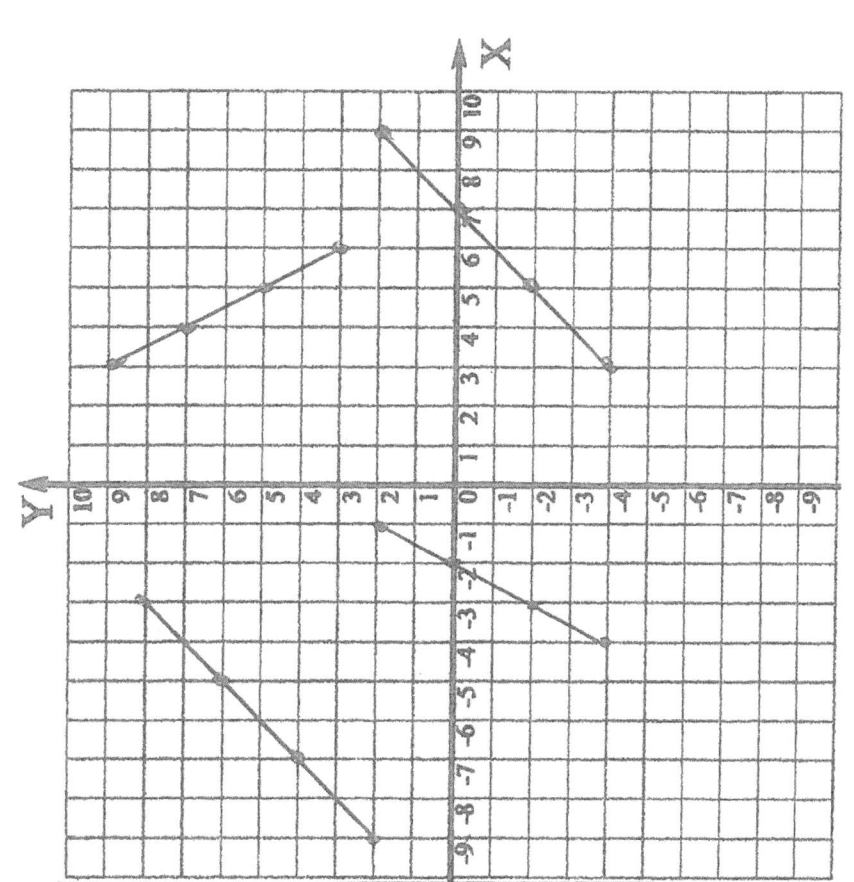

Plot the ordered pairs being called out by putting an X over the point. When you have all four X's along any one line, call out BinGRAPHo!

BINGRAPHO

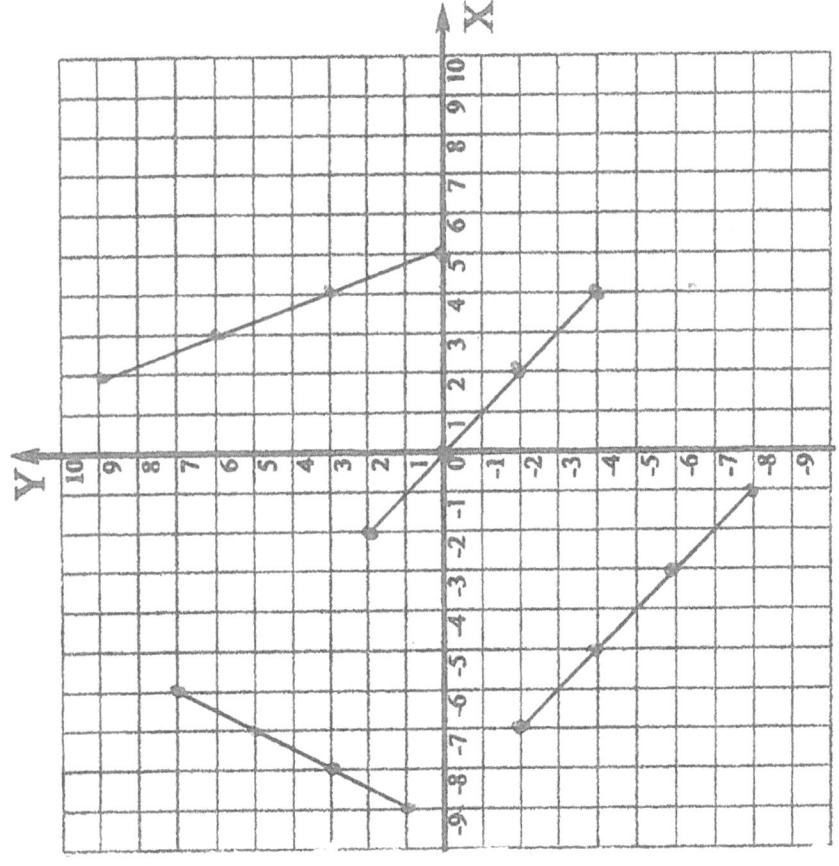

Plot the ordered pairs being called out by putting an X over the point. When you have all four X's along any one line, call out BinGRAPHo!

BINGRAPHO

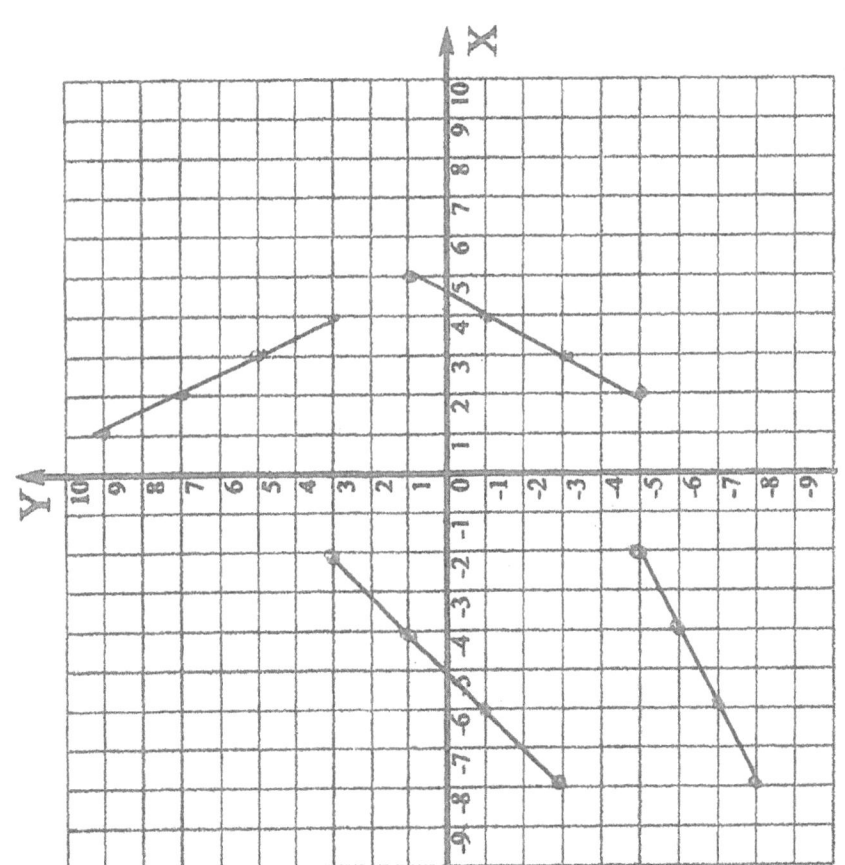

Plot the ordered pairs being called out by putting an X over the point. When you have all four X's along any one line, call out BinGRAPHo!

BINGRAPHO

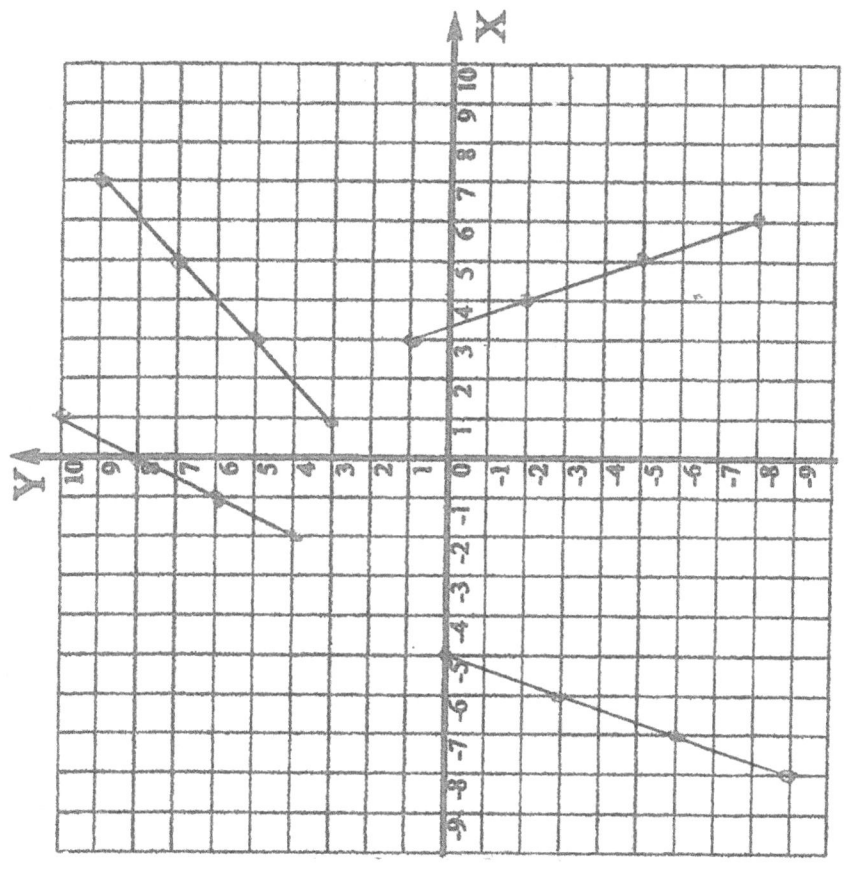

Plot the ordered pairs being called out by putting an X over the point. When you have all four X's along any one line, call out BinGRAPHo!

BINGRAPHO

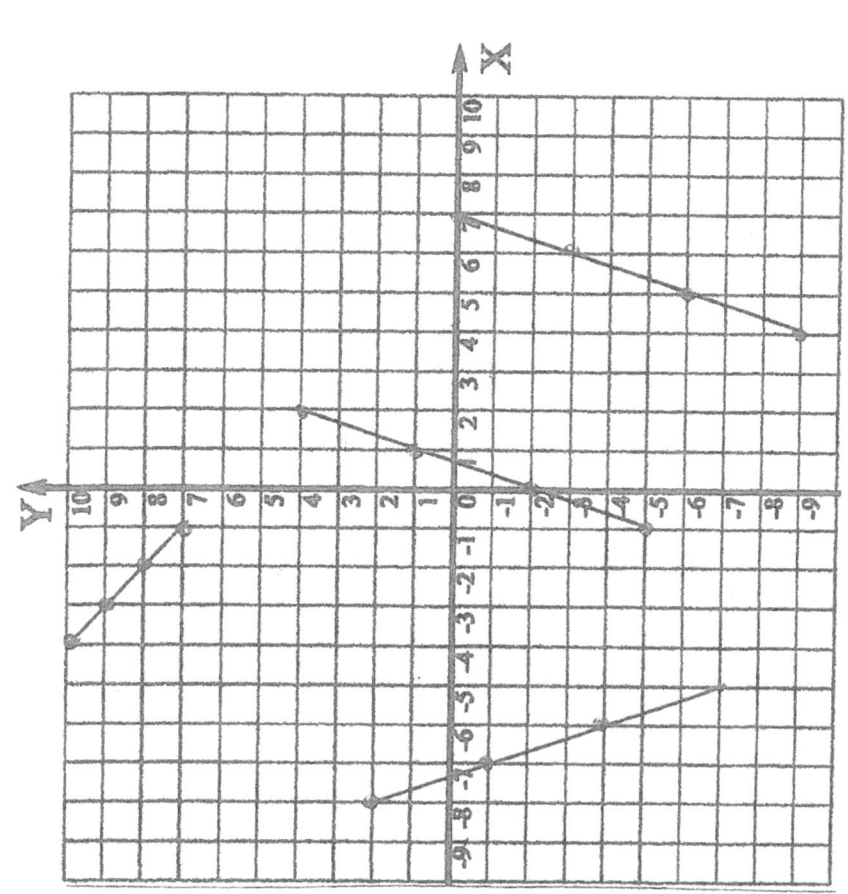

Plot the ordered pairs being called out by putting an X over the point. When you have all four X's along any one line, call out BinGRAPHo!

BINGRAPHO

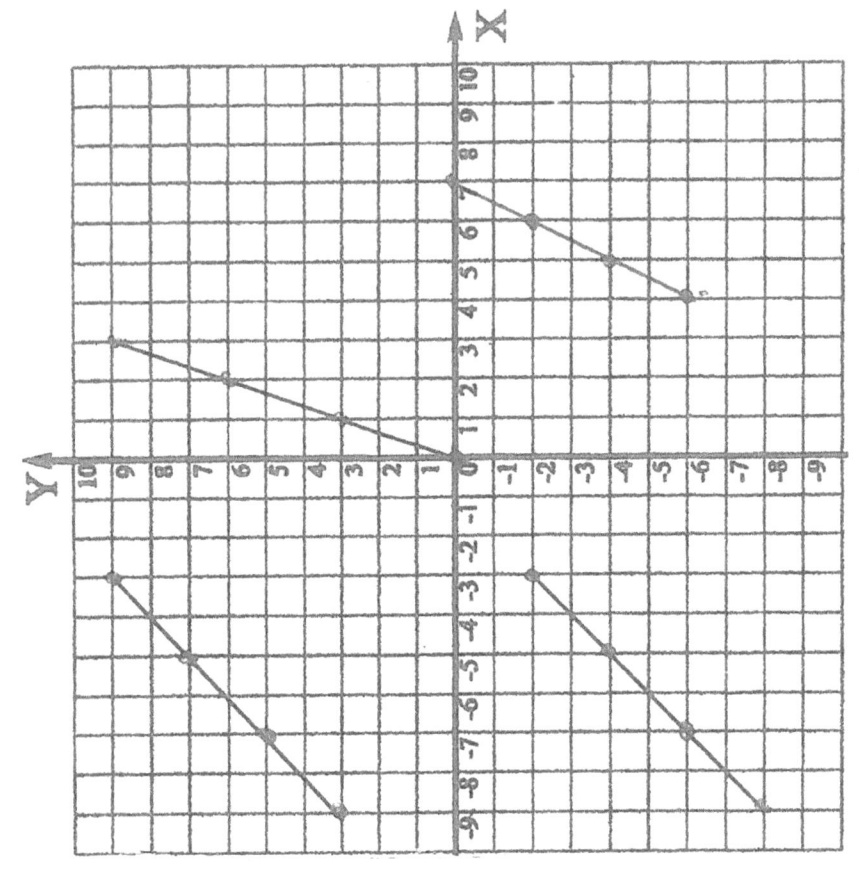

Plot the ordered pairs being called out by putting an X over the point. When you have all four X's along any one line, call out BinGRAPHo!

BINGRAPHO

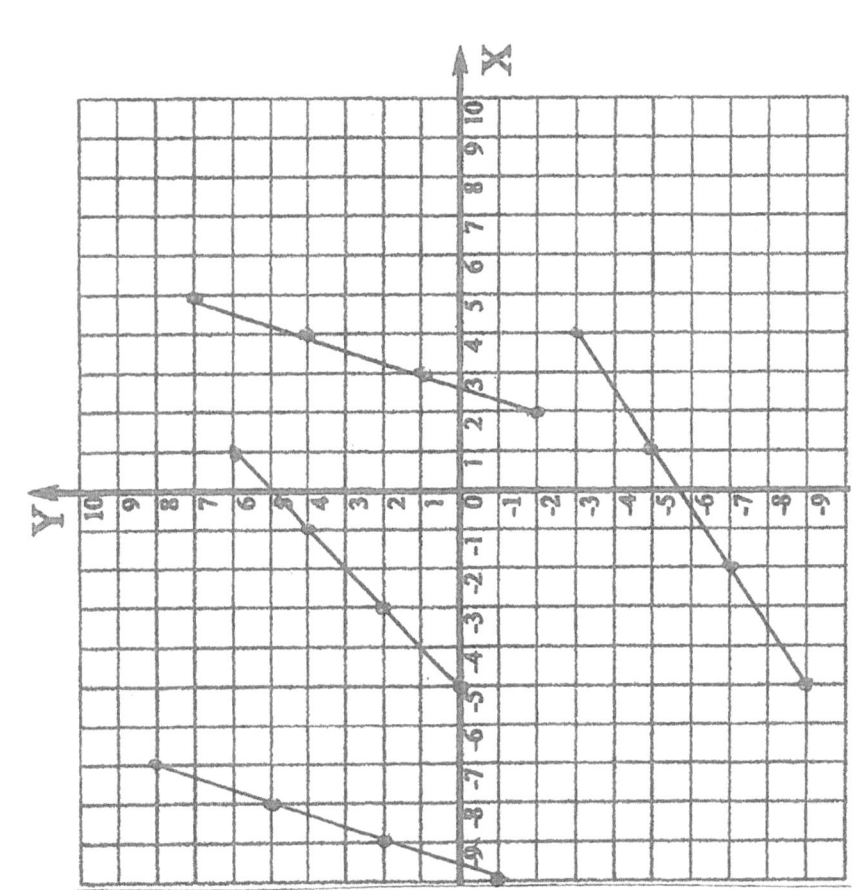

Plot the ordered pairs being called out by putting an X over the point. When you have all four X's along any one line, call out BinGRAPHo!

BINGRAPHO

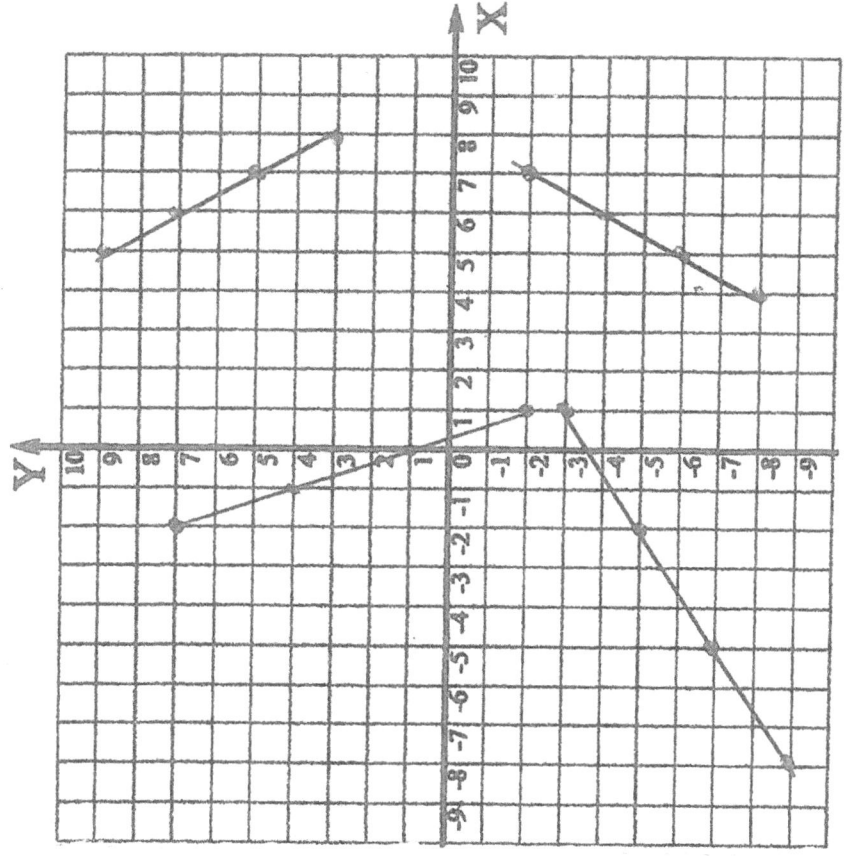

Plot the ordered pairs being called out by putting an X over the point. When you have all four X's along any one line, call out BinGRAPHo!

BINGRAPHO

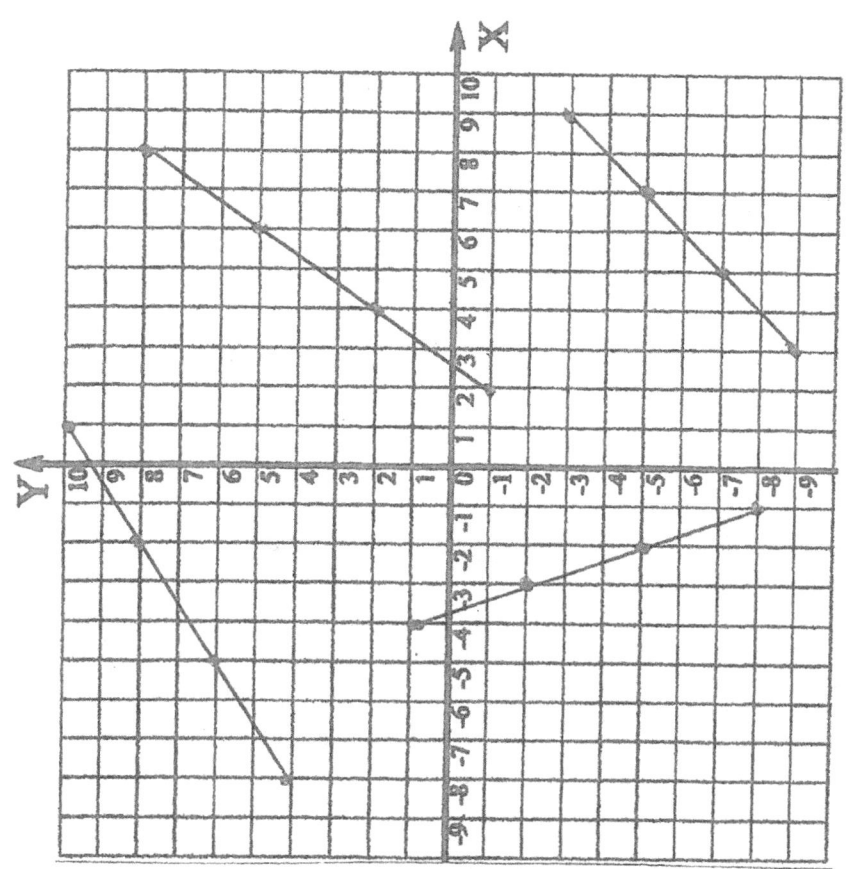

Plot the ordered pairs being called out by putting an X over the point. When you have all four X's along any one line, call out BinGRAPHo!

BINGRAPHO

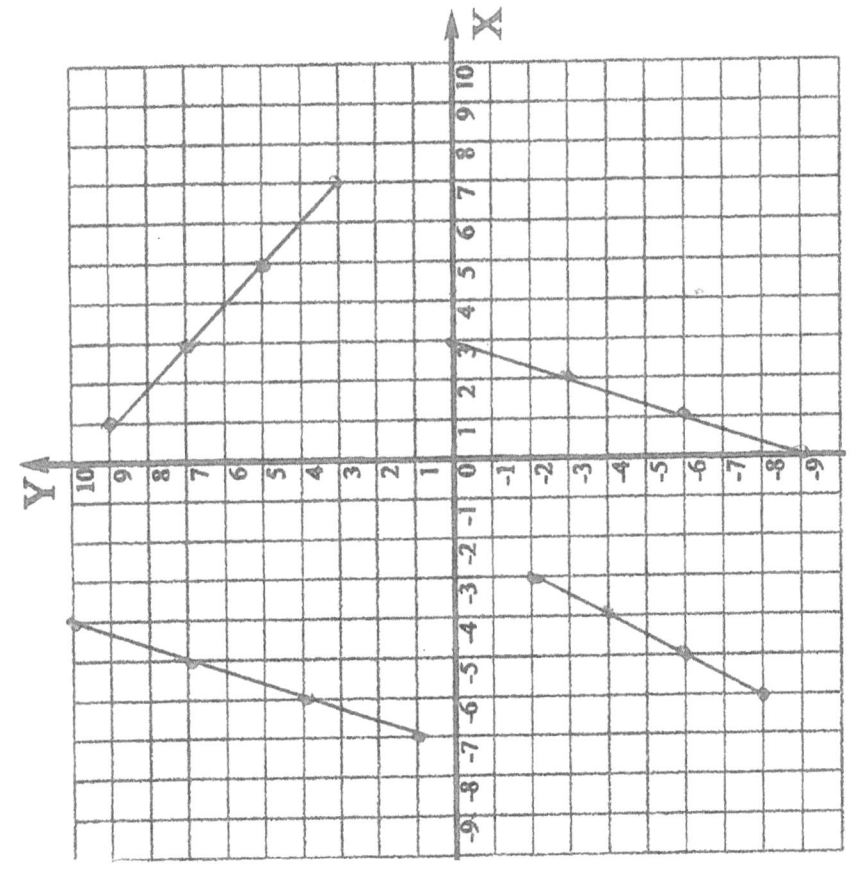

Plot the ordered pairs being called out by putting an X over the point. When you have all four X's along any one line, call out BinGRAPHo!

BINGRAPHO

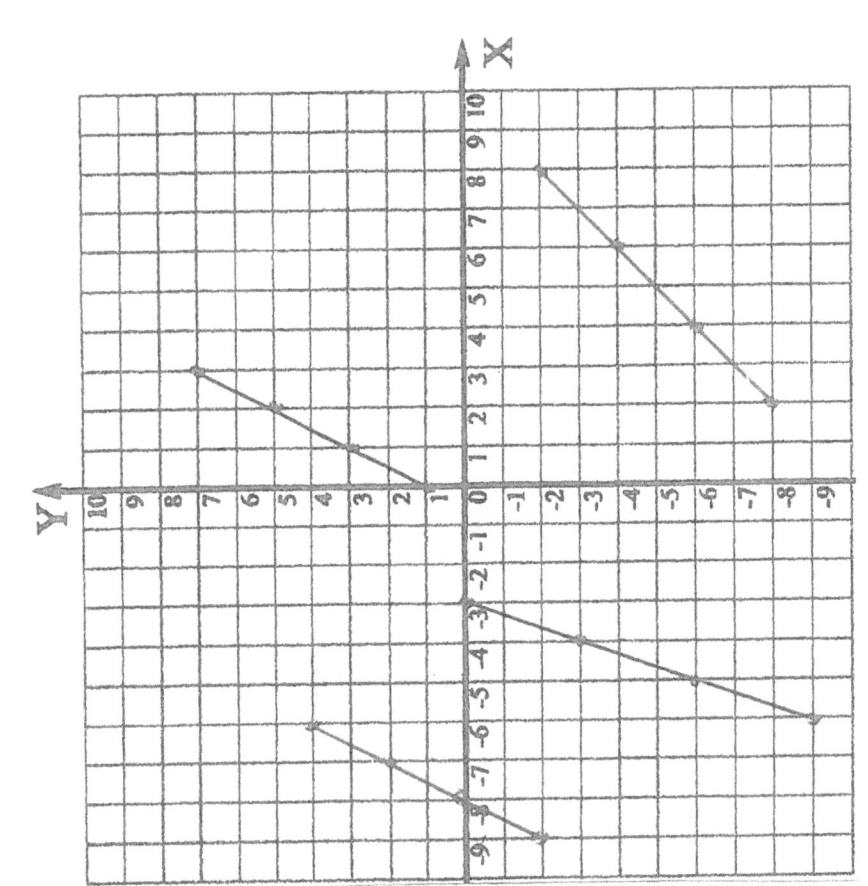

Plot the ordered pairs being called out by putting an X over the point. When you have all four X's along any one line, call out BinGRAPHo!

BinGRAPHo

Answer Sheet 1 of 2

TEACHER RESOURCE:

Ordered pairs for Version Two of BinGRAPHo with the pre-made cards in this book.

(3, 4)	(0, 0)	(-7, 7)	(0, -3)	(4, -5)	(-8, -3)	(2, -2)
(9, 2)	(1, -7)	(-6, 1)	(-6, -1)	(4, -4)	(7, 0)	(2, -11)
(-5, -2)	(-4, 1)	(5, 0)	(5, -2)	(-3, 4)	(-4, -5)	(-2, 3)
(4, 3)	(3, -4)	(-1, 6)	(-3, -8)	(5, 1)	(3, 6)	(-3, 3)
(-1, 2)	(4, -1)	(2, 9)	(-1, 4)	(3, 10)	(2, 4)	(3, -3)
(2, 4)	(1, 5)	(1, 1)	(5, 6)	(2, -5)	(-1, 5)	(3, 6)
(8, 8)	(-2, 2)	(-4, 6)	(-1, 1)	(3, -3)	(-5, 0)	(-3, 2)
(-4, 1)	(-1, 4)	(1, 6)	(4, -3)	(-5, -9)	(-2, -7)	(1, -5)
(-6, -1)	(-2, 4)	(-4, -5)	(10, 3)	(-7, -2)	(-7, 2)	(-1, 7)
(-7, -7)	(-1, -4)	(-8, -5)	(-8, 5)	(1, -3)	(-10, -9)	(1, 8)
(-3, -6)	(-9, -8)	(-9, 8)	(3, -5)	(-8, 0)	(-5, -8)	(0, 2)
(3, 2)	(7, -9)	(-7, 3)	(-7, 10)	(1, 4)	(5, 4)	(-1, 1)
(-5, 9)	(4, 2)	(3, 6)	(7, 6)	(-2, 4)	(-6, 6)	(5, -1)
(9, 8)	(-3, 7)	(1, 9)	(6, -4)	(3, 8)	(-4, -2)	(-4, 10)
(4, 7)	(7, -7)	(-3, -5)	(-3, 1)	(-1, -3)	(7, 5)	(-6, 1)
(-1, -6)	(1, -7)	(3, -8)	(9, 1)	(7, 2)	(5, 3)	(-2, 1)
(-1, -5)	(-7, 2)	(2, -3)	(-2, 2)	(-5, -1)	(0, -7)	(-2, 4)
(-6, 4)	(1, -6)	(-1, -1)	(6, -4)	(-4, -8)	(-1, 6)	(0, 1)

BinGRAPHo Answer Sheet 2 of 2

TEACHER RESOURCE:

Ordered pairs for Version Two of BinGRAPHo with the pre-made cards in this book.

(0, -9)	(0, -4)	(6, -4)	(-4, -8)	(-1, 6)	(0, 1)	(0, -9)
(0, -4)	(3, -6)	(-8, -9)	(0, 8)	(1, 3)	(-8, -5)	(1, -7)
(0, -9)	(-1, 7)	(1, 10)	(2, 5)	(-7, -3)	(1, 0)	(-3, -10)
(-2, 8)	(3, 1)	(3, 7)	(-6, -1)	(0, 3)	(0, -2)	(-3, 9)
(4, -2)	(-7, 1)	(-5, 1)	(-1, 6)	(0, 2)	(-4, 10)	(5, -5)
(-6, 4)	(-1, -1)	(-2, 9)	(-1, 3)	(2, 4)	(6, -8)	(-5, 7)
(0, 1)	(-9, 2)	(-2, 4)	(1, 1)	(-9, -2)	(-4, 10)	(1, 3)
(-7, 4)	(9, 2)	(7, 0)	(5, -2)	(3, -4)	(-3, 8)	(-5, 6)
(-3, 5)	(0, -2)	(-8, 0)	(3, 0)	(2, 5)	(-4, -6)	(-9, 3)
(-7, 5)	(-4, 7)	(-3, 9)	(7, 0)	(6, -2)	(5, -4)	(4, -6)
(-4, 1)	(-3, -2)	(-2, -5)	(-1, -8)	(2, -1)	(4, 2)	(6, 5)
(8, 8)	(1, -2)	(0, 1)	(-1, 4)	(-2, 7)	(5, 9)	(6, 7)
(8, 3)	(1, 6)	(2, 4)	(3, 2)	(4, 0)	(-1, -2)	(1, -4)
(3, -6)	(-9, -3)	(-8, -1)	(-7, 1)	(-6, 3)	(-1, 2)	(0, 0)
(1, -2)	(2, -4)	(-3, -2)	(0, -4)	(3, -6)	(6, -8)	(8, 1)
(6, 3)	(4, 5)	(2, 7)	(3, 2)	(2, 4)	(1, 6)	(0, 8)
(5, -6)	(2, -7)	(-1, -8)	(-5, -1)	(-2, 1)	(1, 3)	(4, 5)
(2, -1)	(1, 3)	(7, 5)	(5, -8)	(8, -5)	(-4, 11)	(2, 6)

BRAIN BUSTING BOARDS

(No Student Worksheet For This Activity)

Create or find/copy a game board on 11" x 17" inch tabloid paper (or any size). Give to each group of students. After each lesson, have students create a problem to put on the game board.

Also have them write something in a space that is FUN and creative. For example: move back $\sqrt{9}$ spaces, or roll again + 2, or stay here until you roll an odd number.

When the board is complete allow students to play. This is a mathabulous way to assess students, or use as a review before an assessment.

Actual Classroom Example:

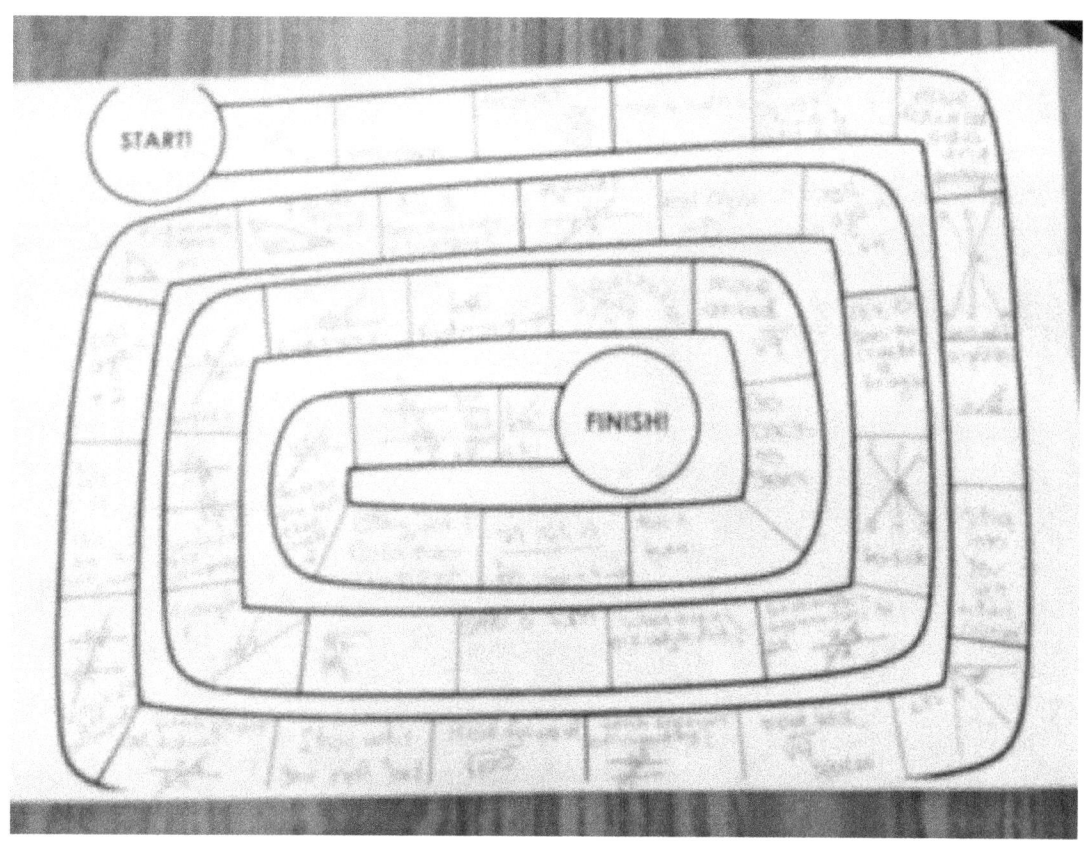

MATHLICIOUS JEOPARDY

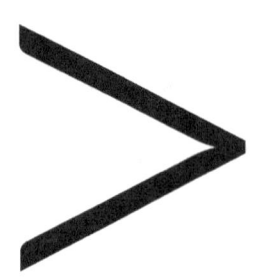

This is a great way to incorporate technology with learning.

www.jeopardylabs.com is a free website which allows students or you to create Jeopardy games with <u>any subject</u>. I put students in pairs or groups to complete this project. They will get a code when they are finished and need to record it or have you write it down. You can put each group's game on the smart board and the class can play. Before you assign the project, go to the website and play a round with your students.

This will take several days to complete.

Mathlicious Jeopardy

Name: _____

Directions: Create your own jeopardy template. You need to create 5 categories and 5 answers with questions under each category. The questions should get a little more challenging as the score increases. Your questions have to match the category. Have fun and be creative. Remember write questions, then the answer in question form ("WHAT IS....?")

Examples: (Q) A triangle with two equal sides.
(A) What is an isosceles triangle?

(Q) The square root of 289.
(A) What is 17?

Category 1: _____

100   (Q) _____
(A) _____

200   (Q) _____
(A) _____

300   (Q) _____
(A) _____

400   (Q) _____
(A) _____

500   (Q) _____
(A) _____

Category 2: _____

100   (Q) _____
(A) _____

200   (Q) _____
(A) _____

300   (Q) _____
(A) _____

400   (Q) _____
(A) _____

500   (Q) _____
(A) _____

Mathlicious Jeopardy

Category 3: _____

- 100 (Q) _____
- (A) _____
- 200 (Q) _____
- (A) _____
- 300 (Q) _____
- (A) _____
- 400 (Q) _____
- (A) _____
- 500 (Q) _____
- (A) _____

Category 4: _____

- 100 (Q) _____
- (A) _____
- 200 (Q) _____
- (A) _____
- 300 (Q) _____
- (A) _____
- 400 (Q) _____
- (A) _____
- 500 (Q) _____
- (A) _____

Category 5: _____

- 100 (Q) _____
- (A) _____
- 200 (Q) _____
- (A) _____
- 300 (Q) _____
- (A) _____
- 400 (Q) _____
- (A) _____
- 500 (Q) _____
- (A) _____

Mathlicious Jeopardy

Name: _____

Category 1	Category 2	Category 3	Category 4	Category 5
100	100	100	100	100
200	200	200	200	200
300	300	300	300	300
400	400	400	400	400
500	500	500	500	500

MORE MATHLICIOUS IDEAS

KEEP BEING MATHLICIOUS!

If you've reached the end of this book, you've probably figured out that when I get started talking about teaching ideas: I can't stop! It is a sickness (that I hope to pass on to you).

I have to pick a stopping point somewhere (unless Volume 2 comes out)! Here are just a few more ideas to keep that positive parabola on your face!

Mathlicious Measuring

I give kids a prism of candy, like a box of Nerds. The first person to compute surface area or volume that matches what you computed wins the full size prism (candy treat)!

You could even have polygons to measure perimeter. The first person to get the perimeter you calculate wins.

Fashizzle and Twizzle

I use pull-n-peel twizzlers to model angles, shapes, quadrants, lines, etc…
Kids love using manipulatives they can eat afterwards.

RhyMATHm

I create math songs with whatever songs are popular for my students. I just change the words into math words.

Examples: It's all about the MATH, bout the MATH, no trouble!
Watch me whip, out this equation. *(We do the Ray Ray instead of nae nae. Instead of stanky leg… it's hypotenuse leg).*

Sometimes I bust out my CeleMATHbration song.

> Celebrate mathematics, come on!
> Bring your fractions your equations and the variables too,
> I'm going to celebrate mathematics with you
> Come on now…Celebrate MATH, come on!!

After singing this, I have students stand up and say a math term. They go around the room and can't repeat anyone else's term. They get about three seconds to think of a word. If they repeat or can't think of any, they sit down. Last person standing wins. I'm telling you it's a celeMATHbration!!

Kids will begin thinking of their own. For example..
> Who let the MATH out???
> I think I found myself a MATH teacher!!

RhyMATHm (cont'd)

My latest is "get vertical on that line" to the tune of JuJu on That Beat. I give the kids tape to create two parallel lines and a transversal on the floor. Then as I sing the song, they put their feet to illustrate the angles I'm calling out.

Walked into the classroom
Students looking at me
Math swag on and heels looking sweet
Hey Hey hey okay okay I want y'all do it, do this dance now
Get vertical on that line get vertical, vertical on that line
Supplementary on that line, supplementary on that line
Now slide drop Hit dem folks don't stop
Corresponding, corresponding on that line
Adjacent, adjacent on that line
Now do your dance, do your dance, do your dance aye
You're smart, you're a math star
Aye Aye do your dance aye
Go crazy, Get freaky aye
Lets go leggo leggo leggo hey hey hey
Okay we computing and solving
And ready to learn
I got my students with me
And Dr. J on the right
And I'm an Illinois baby
And I don't know nothing else
Beside learning, doing math
And having some fun
I say look in the classroom
What do you expect me to do
I see lots of students
And got them beautiful faces
I mean I like your style
I'm on a whole nother dimension
If you compare the love I have for math and students
There wouldn't be no difference
Get vertical on that line, get vertical, vertical on that line
Supplementary on that line, supplementary on that line
Now slide drop Hit dem folks don't stop
Corresponding, corresponding on that line
Adjacent, adjacent on that line
Now do your dance, do your dance, do your dance aye
You're smart, you're a math star

www.ingramcontent.com/pod-product-compliance
Lightning Source LLC
Chambersburg PA
CBHW082332220526
45470CB00008B/2481